The Magic Theorem

The Magic Theorem: a Greatly-Expanded, Much-Abridged Edition of The Symmetries of Things presents a wonderfully unique re-imagining of the classic book, *The Symmetries of Things*. Begun as a standard second edition by the original author team, it changed in scope following the passing of John Conway. This version of the book fulfills the original vision for the project: an elementary introduction to the orbifold signature notation and the theory behind it.

The Magic Theorem features all the material contained in Part I of *The Symmetries of Things*, now redesigned and even more lavishly illustrated, along with new and engaging material suitable for a novice audience. This new book includes hands-on symmetry activities for the home or classroom and an online repository of teaching materials.

John H. Conway was the John von Neumann Chair of Mathematics at Princeton University. He obtained his BA and his PhD from the University of Cambridge (England). He was a prolific mathematician active in the theory of finite groups, knot theory, number theory, combinatorial game theory, and coding theory. He also contributed to many branches of recreational mathematics, notably the invention of the Game of Life.

Heidi Burgiel holds a Ph.D. in Geometry from the University of Washington and a Master's degree from the Harvard Graduate School of Education. Her professional activities range from mathematical fiber arts through computer systems administration. Over the past 30 years she has worked at the University of Washington, the University of Minnesota, the University of Illinois at Chicago, the Boston Museum of Science, Boston University, Bridgewater State University, the Massachusetts Institute of Technology, Harvard University, the University of Massachusetts at Lowell, and Lasell University.

Chaim Goodman-Strauss is Outreach Mathematician at the National Museum of Mathematics (MoMath). Born and raised in Austin, Texas, he earned his Ph.D. in Knot Theory at UT Austin in 1994 and served through 2022 on the mathematics faculty at the University of Arkansas. He has held visiting positions at the Geometry Center at the University of Minnesota, Princeton University, and the Universidad Nacional Autónoma de México.

The Magic Theorem

A Greatly-Expanded, Much-Abridged Edition of The Symmetries of Things

John H. Conway, Heidi Burgiel, and Chaim Goodman-Strauss

CRC Press
Taylor & Francis Group
Boca Raton London New York

CRC Press is an imprint of the
Taylor & Francis Group, an **informa** business

AN A K PETERS BOOK

First edition published 2025
by CRC Press
2385 NW Executive Center Drive, Suite 320, Boca Raton FL 33431

and by CRC Press
4 Park Square, Milton Park, Abingdon, Oxon, OX14 4RN

CRC Press is an imprint of Taylor & Francis Group, LLC

ISBN: 978-1-032-18273-5 (hbk)
ISBN: 978-1-032-16200-3 (pbk)
ISBN: 978-1-003-25374-7 (ebk)

DOI: 10.1201/9781003253747

Typeset in Adobe Garamond
by KnowledgeWorks Global Ltd.

Publisher's note: This book has been prepared from camera-ready copy provided by the authors.

For Product Safety Concerns and Information please contact our EU representative:
GPSR@taylorandfrancis.com.
Taylor & Francis Verlag GmbH, Kaufingerstraße 24, 80331 München, Germany.

This edition is presented as a celebration of
John H. Conway (1937–2020)

Contents

Preface

This book has been brewing for a long time. John Conway was always interested in geometrical groups, and devised particular notations for these when he was teaching at Cambridge University. However, after Bill Thurston shared the orbifold idea in 1985, John dropped those notations forever and devised the signature notation used in this book.

John shared the news with scores of audiences, ranging from the Princeton rug society to the International Congress of Mathematicians. Many remember John rolling around on his back, legs and arms straight in the air, as he demonstrated the symmetry type of a dining room table!

At a presentation to the Smith College Regional Geometry Institute, graduate student Heidi Burgiel took notes for distribution during the conference. Years later, when John spent some time at Northwestern University, Heidi proposed to expand those notes. They soon brought on Chaim Goodman-Strauss, who had been preaching the gospel of the orbifold signature on his own and was known for his beautiful illustrations.

All they had intended to write was the content of what is now this book — an elementary introduction to the orbifold signature notation and the theory beneath it. But then came the idea of writing a second part that would extend the signature to color symmetry. The book continued to grow and grow, until *The Symmetries of Things* (*S.o.T.*) finally appeared in the spring of 2008.

The full edition remains a bargain, with more than three hundred pages of material not replicated here, a wealth of symmetries of things. We ourselves are not just authors, but are also regular users of *S.o.T.*, often consulting its extensive tables and enumerations.

As comprehensive as is *S.o.T.*, we have always felt there is much more to add. Even as the first edition appeared in print, we were preparing material for its revision, with many new examples, applications, and illustrations. That very week, John was playing with "vo-cells" and Chaim unveiled *Double Triamond, w/ Hexastix!*. This sculpture, made with Eugene Sargent, appears below and demonstrates the quarter group $4^o/4$.

Not in the first edition of The Symmetries of Things.

John remained fascinated with the quarter groups and was an inexhaustible well of knowledge and ideas. Heidi continued her work in teacher professional development and art in math, and went on to become an instructional designer. Chaim has continued to produce sculpture, software, graphic work, and hands-on activities showing off the themes of *S.o.T.*

In just the last few years, we have seen a new flowering of geometric ornament, with a flourishing community of designers and mathematical illustrators working with wonderful new software tools alongside ancient craft. Today the ideas we use from topology and geometry are more visible and accessible to mathematicians and non-mathematicians alike, even appearing in widely-known, popular games and videos. The time seems right to say more.

With the passing of our dear friend, teacher, and senior author, the inimitable John H. Conway (1937-2020), a full second edition is too great a task for now.

Nonetheless we feel the Orbifold Theory of Symmetry has yet to come completely into its own. As magnificent, if we may say so, as is the full edition of *The Symmetries of Things*, its very solidity (and expense) has kept its message from reaching as wide an audience as it could. In this new book,

The Magic Theorem:
a greatly-expanded,
much-abridged edition of
The Symmetries of Things

we aim to bring this notation to the appreciation of a wider audience, presenting orbifolds and the magic theorems that utilize them for anyone to enjoy, in a slimmer, more accessible (and cheaper) volume.

This new book is essentially Part One of *S.o.T.*, closer to our initial vision, redesigned and lavishly illustrated, with quite a lot of new and additional material such as:

- A new Chapter 9 on the orbifold theory which may be read alongside the rest of this book.

- Many new examples and exercises, as well as their solutions.

- A large number of photographs of symmetrical things designed by other artists and artisans, contemporary and classic.

- Many extra images prepared for *S.o.T.*, unseen until now.

- Much new artwork prepared since.

- Hands-on symmetry activities for the home, classroom, or office, available at *themagictheorem.com*, listed on page 168, including:

  ○ a large number of polyhedra to download and assemble,

  ○ even more examples of geometric ornament to analyze,

  ○ and lots of things to create with paper, scissors, and tape.

- A supplementary Chapter 10 on hyperbolic geometry and Thurston's fullest form of the Magic Theorem.

We're especially pleased to show you so many repeating patterns that we have found in the world around us, and we know that you will enjoy finding and analyzing the repeating patterns all around you too.

We hope this book will help inspire a new generation of mathematicians, designers, scientists, artists, and educators to use orbifolds to explore the wonders of symmetry!

Acknowledgements

The Symmetries of Things and *The Magic Theorem* would not have appeared without the help of many people including Javier Bracho, Vladimir Bulatov, Marc Culler, Sandy Dirkx, Peter Doyle, Callum Fraser, Olaf Delgado Friedrichs, Sarah Garvin, Troy Gilbert, Edmund Harriss, Charlotte Henderson, Tracy Hicks, Keith Hollingshead, Daniel Huson, Natasha Jonaska, Cindy Lawrence, Silvio Levy, Sabetta Matsumoto, Tom Moore, John Overdeck, Alice and Klaus Peters, Siobhan Roberts, Saul Rosenthal, Tom Rogers, Eugene Sargent, Doris Schattschneider, Henry Segerman, Marjorie Senechal, Neil Sloane, Peter Stampfli, Robert Strauss, Bill Thurston, Jade Vinson, Jeff Weeks, Lawrence Weschler, Chris Whatley, the very many readers who sent us corrections from the full edition, our friends, students, and colleagues, and the patience and sympathy of the authors' partners Diana and Kendall.

We are especially appreciative of all those who shared their artwork for this book (listed on page 169), and thank the institutions that supported us, including Bridgewater State University, Lasell University, the National Museum of Mathematics, the National Science Foundation, Northwestern University, Princeton University, the Universidad Nacional Autómata de México, the University of Arkansas, and the University of Illinois at Chicago.

Symmetrically yours,
Heidi and Chaim,
October 2024.

Introduction

Symmetries and symmetric patterns surround us throughout our lives. The aim of the first half of this book is to describe and enumerate all the symmetries found in repeating patterns on surfaces. To prove that our enumeration is accurate, we then explain the beautiful ideas from topology and algebra which form the basis for our conclusions.

We start with a problem – enumerating symmetric patterns.

We then introduce tools for solving this problem and complete the enumeration. But then we are presented with a second problem – demonstrating that these tools work the way we claim, that there is a solid mathematical foundation beneath our results. Again, we solve this problem with some tools, then present the mathematics supporting the use of those tools. In this way, each chapter reduces the problems left by the preceding chapter to another problem whose solution is postponed to the following chapter.

This is a departure from the traditional practice of building a theory starting with basic principles and working toward the ultimate goal of proving some final theorem. We believe that our backward approach will be successful because it allows us to present one concept at a time, at the cost of always postponing the proof of just one thing to the next chapter. We hope also that the argument will be clearer when presented in a single logical thread, of the form $A \Leftarrow B \Leftarrow C \Leftarrow ... \Leftarrow Z$.

We suggest you read the first few chapters of this book with markers in hand to analyze its repeating patterns for yourself. (A real mirror or two will certainly be helpful for spotting kaleidoscopic symmetries.)

The first chapter is a gentle introduction to symmetry. Chapter 2 introduces the four fundamental features that we use to classify symmetry. In Chapter 3 we state our Magic Theorem, and apply it to find the 17 possible types of repeating planar patterns, while Chapters 4 and 5 perform a similar service for spherical and frieze patterns.

In Chapter 6, we turn to the topological tools that underly the theory: The Magic Theorem is deduced in Chapter 6 from Euler's Theorem, which is itself proved in Chapter 7. Chapter 8 gives our ZIP proof of the classification of surfaces, completing the proof of the Magic Theorem.

In Chapter 9 we show many orbifolds of the patterns in this book and how to make your own. You may read this chapter alongside the earlier ones to better understand how our fundamental features are used to describe the orbifold of a pattern.

Finally in Chapter 10, we summarize the general situation: the magic theorems for the planar, spherical, and frieze patterns are special cases. Essentially every orbifold symbol corresponds to the symmetry type of a repeating pattern, generally one in the hyperbolic plane.

But what is an orbifold and why are they useful to the study of repeating patterns?

The Orbifold Theory of Symmetry

The orbifold perspective provides a modern and complete mathematical theory of planar symmetry.

Every symmetrical pattern like the ones that you will find throughout this book and in the world around you is associated with a particular surface, its *orbifold*, which we imagine by considering points of the same kind to actually *be* the same point. For example, folding over and fusing the two halves of the shape at left gives us its orbifold, the half-heart shape at right.

Our notation records features of the topology of the orbifold — this orbifold is a surface with a boundary along a fold line, which we denote with a *. In fact, it is these topological features that we will learn to recognize as we master the orbifold notation. You can see many examples of these orbifolds in Chapter 9.

The magic theorems we will use to understand symmetrical patterns rely on powerful and direct tools from the study of the topology of surfaces — Euler characteristic (Chapter 7), the Classification of Surfaces (Chapter 8), and in the most general case, the Gauss-Bonet Theorem.

The key insight is that the features of a pattern's orbifold determine its symmetry type. These tools constrain this topology tightly, with a single number, the "cost" of a symmetry type. Our magic theorems then simply list out the symmetry types with a suitable cost.

Why Orbifolds?

- The Magic Theorems reduce the problem of listing out the possible symmetry types of repeating patterns to a simple arithmetic calculation, the "cost" of each type.

- The theory extends uniformly across patterns in the plane (those with a "cost" of $^\$2$), on the sphere (costing less than $^\$2$), and even those in the hyperbolic plane (costing more). Frieze patterns are simply those with cone or gyration points of infinite order.

- The notation is well-defined and unambiguous, as it records topological features of a pattern's underlying orbifold, and these determine the symmetry type. (See page 143 for the illuminating example of **22×**.)

- Each symbol gives a presentation of its symmetry group (Chapter 15 of *S.o.T.*).

- We can determine when and exactly how a pattern can be stretched or skewed without changing its symmetry type (Chapter 18 of *S.o.T.*).

- We can use the topology of orbifolds to enumerate various types of patterns. For example, in Chapter 16 of *S.o.T.* we list out the tilings that are formed from a single tile moved about by a symmetry, by drawing directly on the orbifold. These tilings are the essentially the same for signatures of the same typographical form. In a similar manner, in Chapter 19 of *S.o.T.* we show how to enumerate all Archimedean tilings working from the topology of their underlying orbifolds.

- To find the symmetry type of a pattern, you need only find its orbifold!

Chapter 1

Symmetries

Every day we are surrounded by symmetric objects and patterns. From furniture to flooring, symmetry is the rule. In art, symmetry is pleasing to the eye, and the intricacies of extremely symmetric patterns can entrance an audience. In architecture, symmetric designs are attractive for yet another reason — repetition of a design element means re-use, which ultimately requires less planning and testing. In manufacturing, it is simpler, cheaper, and more efficient to repeat a pattern at regular intervals. Even Nature has reasons to use symmetry in her work.

Recently, John H. Conway and William Thurston adapted Murray MacBeath's mathematical language for discussing symmetry. Now, the symmetries of a pattern can be defined by a single symbol that we call its signature: for example, 3*3, for the pattern on the previous page. With some practice, anyone can read this signature and identify the symmetries it describes. You'll soon recognize the signatures of the everyday things around you, and perhaps you will wonder how many different types of symmetry there are.

Together in this book, we'll learn to recognize different kinds of symmetries and how to find their signatures. Using the Magic Theorem we will work out all of the types of symmetry that it is possible for a planar pattern to have. In the second half of this book we'll show how each symmetry is related to a special surface, its *orbifold*, and use powerful tools from topology to prove the Magic Theorem. As you read this book, you can turn to Chapter 9 to see examples of orbifolds of many of the patterns we will encounter.

The word *symmetry* is a combination of the words *sym* (together) and *metron* (measuring). The *symmetries of a thing* are those transformations — such as rotations or reflections, even doing nothing at all — that do not change its shape or size.

The *triskelion* above appears on the coat of arms of the Isle of Man and looks the same in three orientations; the rotation through 120 degrees is a congruence that takes the figure to itself. A triskelion has order 3 gyrational point symmetry and signature 3•. The pattern on the previous page — which to a mathematician extends forever in every direction! — has reflections and gyrations, and signature 3*3.

Kaleidoscopes

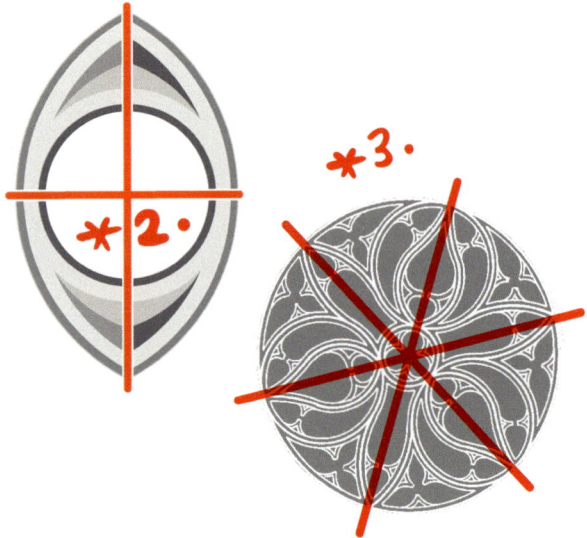

The simplest signature is just ✻ (star). A ✻ denotes a *mirror* or *kaleidoscopic* symmetry, and a ✻ alone means that there are no other symmetries to the figure. This pair of gryphons has a single line of mirror symmetry running between them. Reflecting the image across this mirror line wouldn't change its appearance.

The *vesica piscis* ("fish bladder"), at left, has signature ✻2•, pronounced "star two point symmetry" or, more formally, "order two kaleidoscopic symmetry fixing a point." We use stars for kaleidoscopes to suggest the star formed by the mirrors through a kaleidoscopic point. The *order* of a kaleidoscopic point is the number of mirror lines through it. In this case two lines of mirror symmetry — one vertical, the other horizontal — meet at the center of the pattern. Finally, the point (•) indicates that all the symmetries fix a point, the order two kaleidoscopic point where the mirrors cross.

You can probably guess that in a figure with signature ✻3•, three lines of mirror symmetry meet at its center and similarly for signatures ✻4•, ✻5•, ✻6•, and so on.

Find some signatures: Mark the mirror lines and find the signatures in these stone traceries. You can check your answers on page 11.

To find mirror lines in a pattern, it often helps to use a real mirror, or even a shiny dinner knife. When your mirror is upright on a mirror line, the reflected image on the mirror will match up with the real pattern behind it. The mirror will seem to disappear, just like the ones in the photographs on this page.

QUIZ: *Identify the mirror lines and signatures of these cut-paper snowflakes.* (Answers are on page 11.)

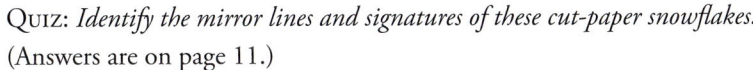

Cut-paper snowflakes like these are made by folding paper and cutting through the layers. When we unfold a snowflake, our cuts in the paper are the same on both sides of each fold line.

These fold lines are exactly the mirror lines in the pattern — you can turn to page 132 to learn more. At right we've folded paper with four mirror lines, and this snowflake has signature ∗4 •. In the photograph, different ways of folding produced patterns with different signatures.

Fold up paper to cut your own snowflakes!

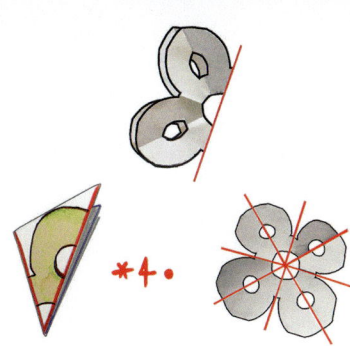

Gyrations

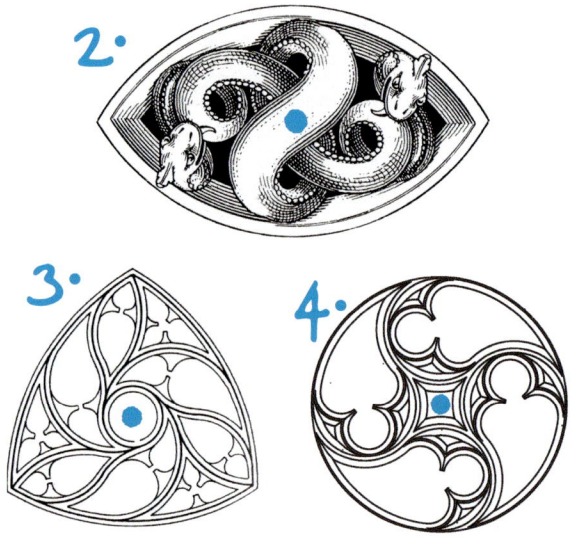

This snake design has no kaleidoscopic symmetry yet looks the same in two orientations: leaving it as is or rotating through 180 degrees, congruences that take the figure to itself. This design has 2-fold *gyrational symmetry*, centered on a 2-fold *gyration point*, and signature 2•.

The stone tracery design at far left has no kaleidoscopic symmetry either, yet it looks the same in three orientations: a rotation through 120 degrees in either direction is a congruence that takes the figure to itself. This tracery has order 3 gyrational point symmetry and signature 3•. The tracery at near left has signature 4•.

QUIZ: *These hubcaps have gyrational symmetries. What are their signatures? Do they change if you take into account the brand logos?* The answers are on page 11.

Spin your pattern to test for gyrations! Here we placed the tip of our pencil at the center of this pattern. We spun the pattern around 1/2 of a revolution but the pattern looks the same, as though it hadn't moved at all. There aren't any mirror lines, and this isn't a kaleidoscopic symmetry. We are spinning around a gyration point, which we mark with the pattern's signature **2•**.

Rosette Patterns

Obviously, we could keep going like this, generating pictures with order 37 kaleidoscopic point symmetry or order 42 gyrational point symmetry. But what else can we do?

For the finite *rosette patterns* like those on the last few pages, there are no other signatures. In a finite pattern, all symmetries of the pattern must fix (i.e., cannot move) the center of the pattern. Reflections across the center of the rosette and rotations about its center are the only symmetries that do this, so they're the only symmetries such a pattern can have.

By experimenting with different combinations of rotational and reflective symmetries, you can easily convince yourself that the types **∗•**, **∗2•**, **∗3•**, **∗4•**, ..., **∗N•** and **2•**, **3•**, **4•**, **5•**, ..., **N•** are the only signatures possible for rosettes, to which we add **1 •** (or just **•**) for the type with that has only one symmetry: leaving a pattern alone.

QUIZ: *Some of these Gothic tracery patterns have mirror lines and a kaleidoscopic signature. Some of them only have gyrational symmetry. Which are which, and what are their signatures?* You can check your answers on page 11.

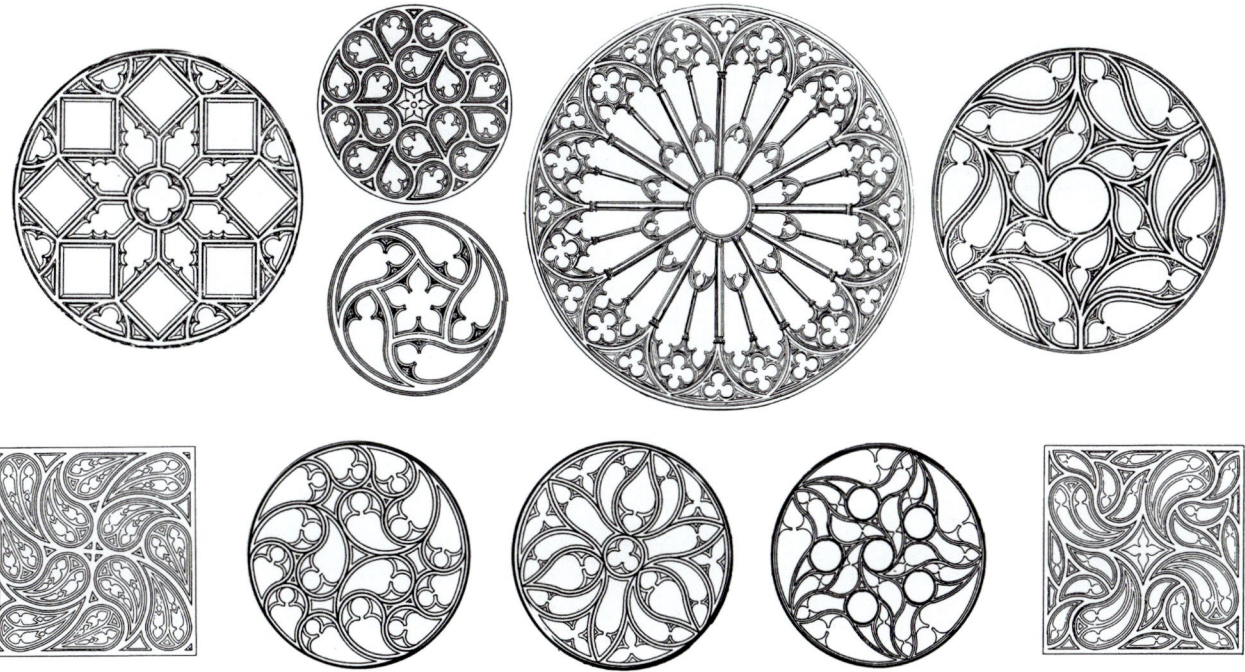

QUIZ: *These rosette patterns were spotted around town. Which of these are kaleidoscopic and which are gyrational? What are their signatures? (See page 11 for the answers.)*

QUIZ: *From A to Z what are the signatures of the letters of the alphabet?*

Frieze Patterns

In the rest of this chapter, we show you the kinds of patterns that we will analyze later in the book. After isolated pictures on a page, the easiest patterns to understand are those made by repeating pictures in a row. We see patterns like this in friezes, ribbons, animal tracks, and fences, even as ancient signature seals.

In addition to any reflective and rotational symmetries of the figures that make up the pattern, a frieze pattern has a translational symmetry that takes the figure to a neighboring figure. This book concerns itself with patterns of this sort, called *repeating patterns*. We'll learn how to analyze frieze patterns in Chapter 5. (Turn to page 94 for the signatures of the patterns below.)

Repeating Patterns on the Plane and Sphere

Frieze patterns have "forward and back" translational symmetry. Plane patterns add translational symmetry in another direction. These patterns can extend to cover an entire page, or even the infinite Euclidean plane. We see them every day on the floors and walls around us. The next few chapters describe how to identify all of the symmetries of these patterns. We've put the symmetry types of these patterns on page 57 if you'd like to see the answers now.

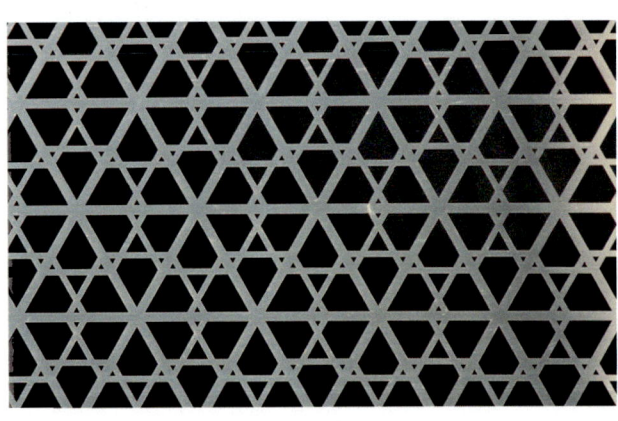

In order to study the symmetries of common objects like hairbrushes and furniture, we will also need to learn about the symmetries of patterns on spheres, because the symmetries of an object will always be the same as those of a sphere wrapped around it. In Chapter 4 we also study things with more elaborate spherical symmetries such as these:

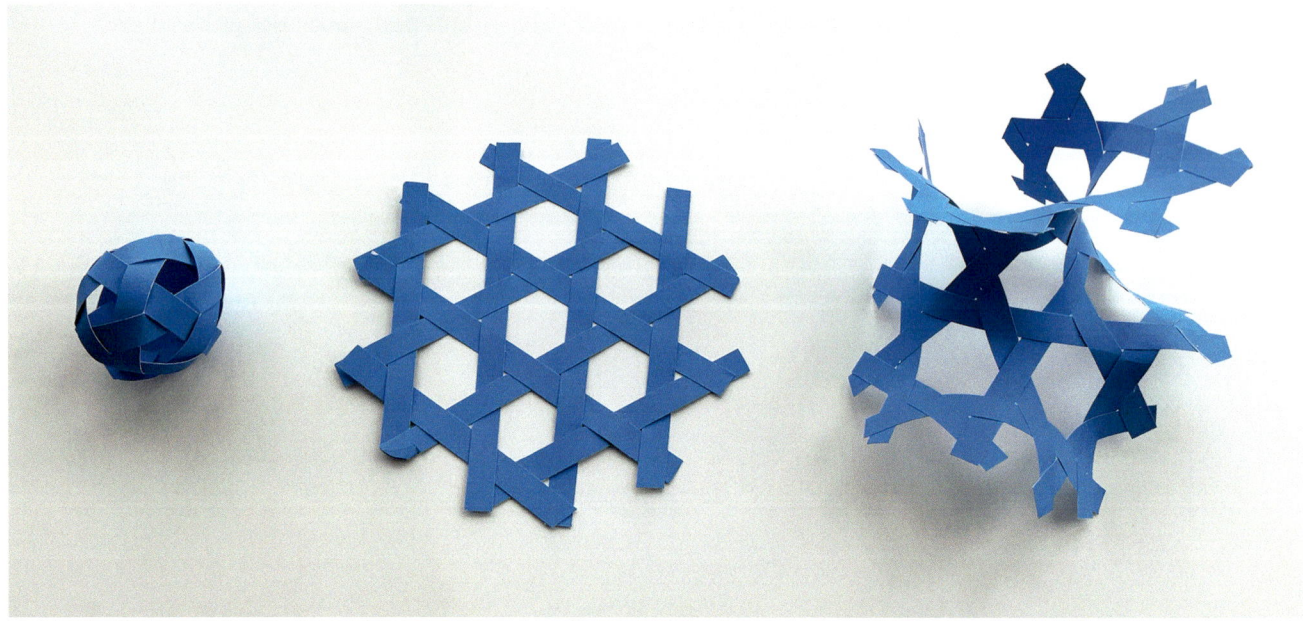

Woven patterns on the sphere, plane and hyperbolic plane with signatures **532**, **632**, *and* **732** *respectively.*

Where Are We?

At the beginning of this chapter we described all of the possible types of symmetry for rosettes — namely • = **1**•, **∗**• = **∗1**•, **2**•, **∗2**•, **3**•, **∗3**•, **4**•, **∗4**•, …. We've also introduced three categories of repeating pattern — frieze patterns, repeating patterns in the Euclidean plane, and patterns on the sphere. This book classifies the different types of symmetry that objects in these categories can have. We've told you roughly what it means to say that two things have the same type of symmetry, but we'll have to postpone a precise definition of our problem until we've nearly solved it.

In fact, our book will have about as many postponements as chapters! For example, in the next chapter we'll introduce four features that are enough to completely specify symmetry types, but will postpone the proof that they do so. These features determine the signatures that we use in Chapters 2–5 to list all possible types for the kinds of patterns we've shown in this chapter. To do so, we employ a "Magic Theorem" whose proof is postponed to Chapter 6.

In that chapter we see that the signature describes a topological surface, an *orbifold*, that encapsulates all the symmetries of a pattern. The Magic Theorem is simply a statement about a number, the Euler characteristic, that describes the topology of the orbifold. In turn, we postpone detailed investigation of the Euler characteristic until Chapter 7.

In Chapter 8 we learn that Euler's characteristic really does characterize the different possible topological types of surface. Our "zip proof" of this will then close up proof of the Magic Theorem. In Chapter 9 we show you orbifolds of many of the patterns in this book.

The orbifold theory we set out is simple and natural. It explains and describes all the different kinds of pattern we've seen in this chapter and, as we'll conclude in Chapter 10, smoothly handles patterns in the hyperbolic plane too.

Answers for Chapter 1

The traceries of page 2 have kaleidoscopic symmetry types

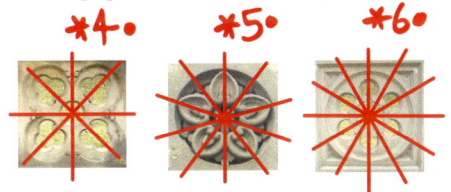

Here are the signatures of the paper snowflake symmetries shown on page 3. Patterns to print and cut out are in the online supplement (pg. 168).

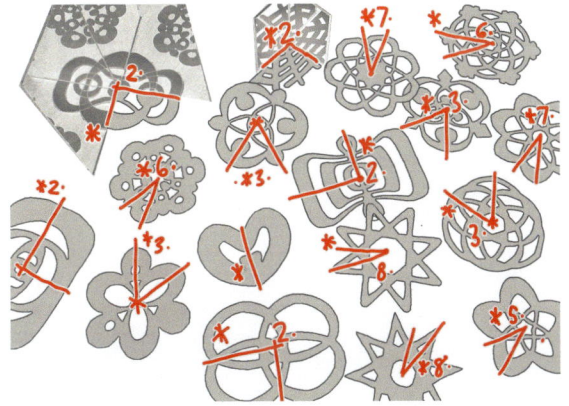

None of the hubcaps on page 4 have mirror lines; all of them have only gyrational symmetry. We've marked the signatures of the hubcaps without considering logos or bolts at their centers. For example, the hubcap at top left has signature **21•** if you don't consider the pattern of bolts at its center; if you do, the hubcap has signature **3•**. If we consider all their details, most of the others have only the "trivial" symmetry of doing nothing at all, with signature we write as **1•** or simply **•**. Many hubcaps in the world have kaleidoscopic symmetry — look around to find them!

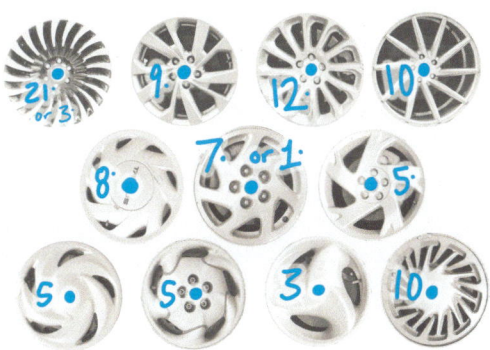

On page 6 we asked for the symmetries of the letters in the Roman alphabet. Many letters have mirror symmetry, or approximately so! (Symmetry will vary from typeface to typeface.) The letters WAVYTUM and BDECK have a mirror symmetry and signature **∗•**. The letters HIX have two mirrors meeting at their centers and signature **∗2•**, and the same for an oval-shaped O. The letters SNZ have gyrational symmetry and symmetry with signature **2•**. The letters FGJLPQR have only trivial symmetry, with signature **•**.

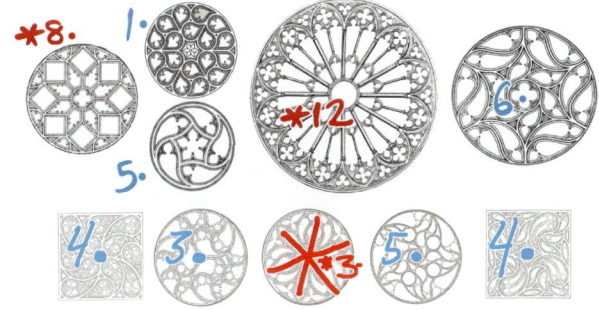

At the top of page 5, many of the traceries have elements with different symmetries overlaid on top of one another. Paying attention just to the main design, the signatures, from left to right, top row, are **∗8•**, **•**, **5•**, **∗12•**, and **6•**. On the second row are **4•**, **3•**, **∗3•**, **5•**, and **4•**.

Below are signatures of some rosettes that we spotted around town. Alone, the peace sign has signature **∗•**, as does the heart within it. Together they have trivial symmetry **•**.

The next few chapters will explain how to find the signatures of repeating patterns. After you've had some practice, you can work out the signatures on the rest of the patterns in this chapter, and check your answers on pages 57, 82, and 94.

Chapter 2

Planar Patterns

This book helps you understand the symmetries of things. In this chapter we look at some repeating patterns and introduce you to the way we think about them. We describe the four fundamental features of a repeating pattern in the plane (or on any surface!) and introduce the signature we use to record them. These features describe the underlying topology of a special surface associated with a symmetry type, its "orbifold" — many examples of these are shown in Chapter 9.

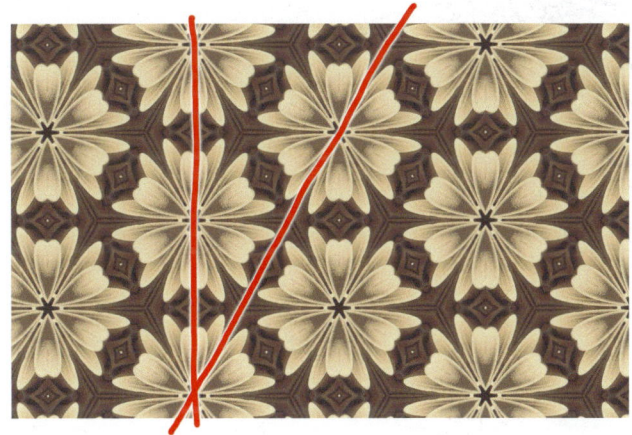

Mirror Lines

The floral pattern to the left has many symmetries. For example, the pattern is left-right symmetric: It has the vertical *mirror line* shown on the figure to the right, and many more vertical mirror lines besides.

We've also drawn another mirror line, which is a different kind because, unlike the first one, it runs *between*, rather than *along*, the petals. Draw some more of the mirror lines on the pattern — and then turn the page to see if you have found them all!

Place three mirrors around the triangle at top above, and look inside to see the floral pattern at left! The mirror lines around the triangle together form the kaleidoscope ∗632, which is the signature of the symmetry of this pattern.

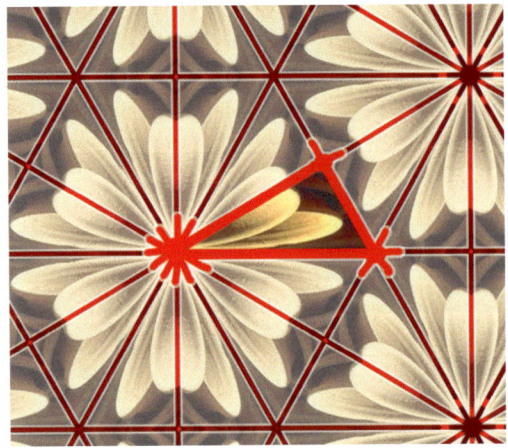

Drawing all the mirror lines we can, we get something like the figure to the left, which is at first sight rather confusing. Fortunately, the small part we've highlighted — and any other part like it, such as the triangle at the top of this chapter — contains enough information to reconstruct the whole pattern.

This is because if we surround the small triangle by mirrors, as in a kaleidoscope, the reflections of the original triangle will fill in the neighboring triangular regions. The reflections of these reflections will fill in the neighbors of these neighbors, and so on, until the entire pattern is restored.

With three small pieces of mirror and a little dexterity, you can try this yourself!

Finding Kaleidoscopes

Patterns whose symmetries are defined by reflections are called *kaleidoscopic* because of their similarity to the patterns seen in kaleidoscopes. They are classified by the way their lines of mirror symmetry intersect.

So, for instance, in these figures there are three particularly interesting kinds of point, shown below: one where six mirrors meet, one where three mirrors meet, and one where two mirrors meet. We call these 6-fold, 3-fold, and 2-fold kaleidoscopic points, respectively, because the local symmetries are *6•, *3•, and *2•.

The mirror lines chain together to form a *kaleidoscope*, which we mark as *. (There are infinitely many repeated copies of this kaleidoscope. We highlight one to look at.)

In this pattern the kaleidoscope is a triangle, and its corners are one of each of a 6-fold, a 3-fold, and a 2-fold kaleidoscopic point. As we shall see, the type of symmetry of the pattern is determined by the type of this kaleidoscope — we indicate this type by its *signature*. This kaleidoscope and this pattern have signature ***632**

The numbers in the signature of a kaleidoscope can be cyclically permuted, so that *632, *326, and *263 mean the same, or also reversed, equating these with *236, *362, and *623. We'll put our digits in decreasing order when we can; hence for this signature we write *632. Turn to page 133 to learn more about the orbifolds of patterns with this signature.

The patterns on this page are somewhat simpler, but if we mark the mirror lines, we see that they have all of the symmetries of the first pattern, and the same kaleidoscope, with signature *632. With a mirror, you can find more mirror lines and check that the kaleidoscopes that we've drawn replicate the patterns.

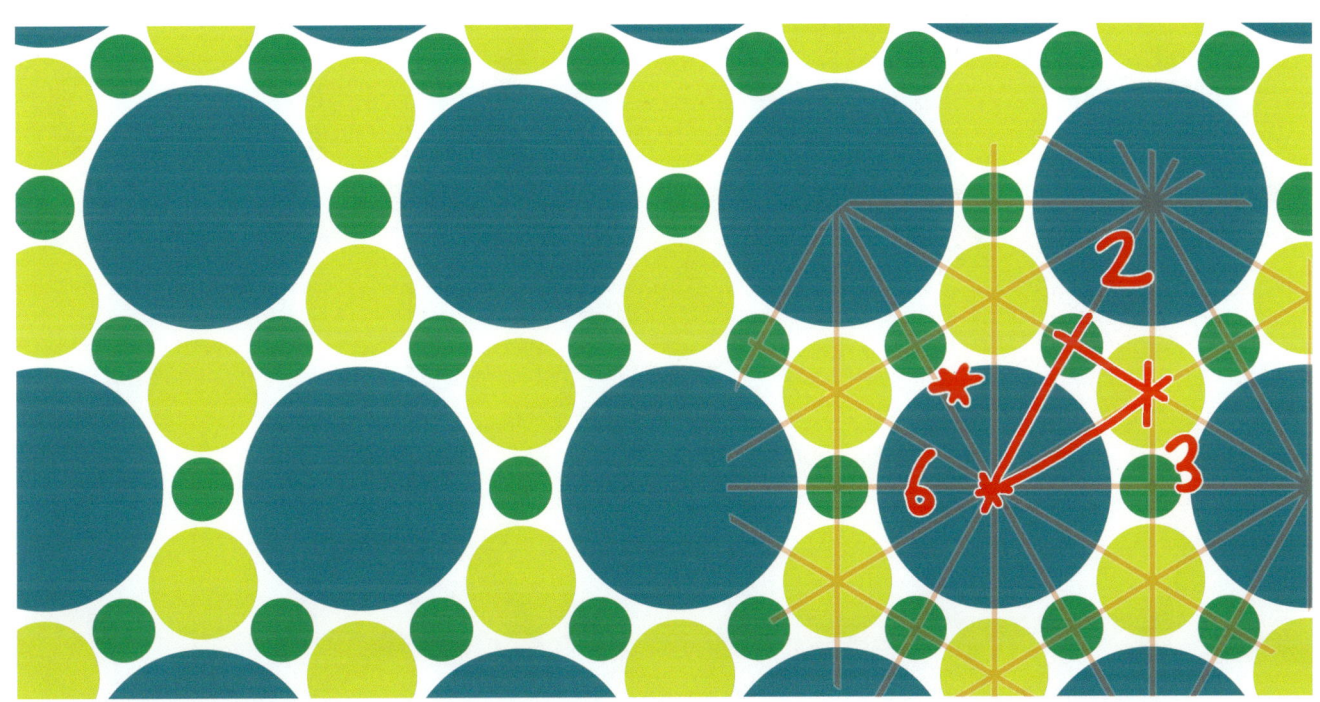

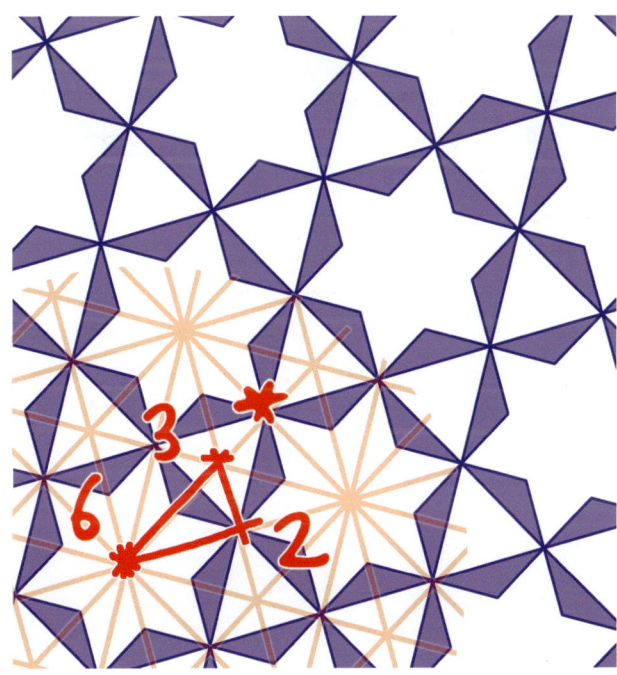

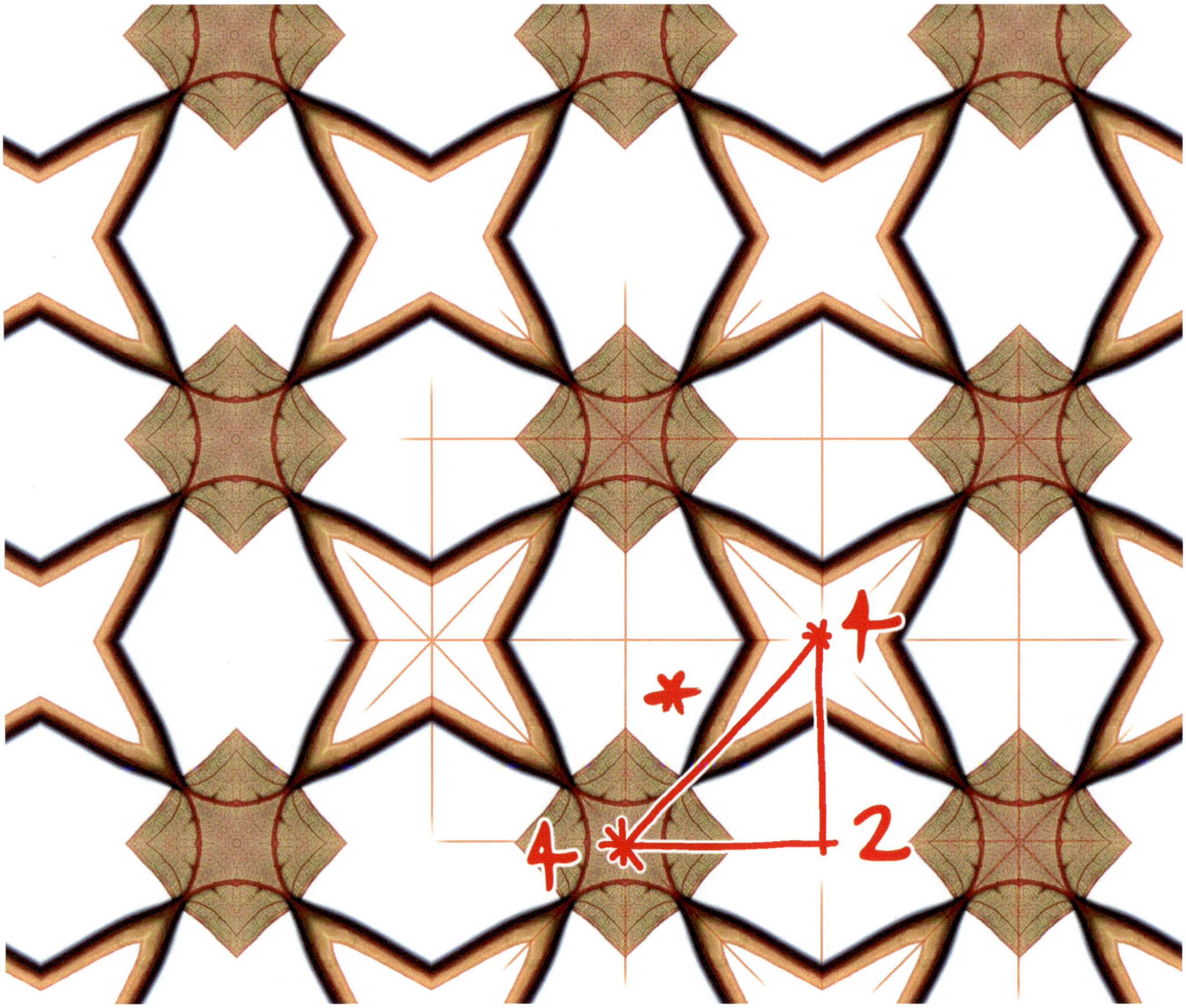

Patterns with a squarish sort of symmmetry such as this one are more common. Marking all of the mirror lines, we find the kaleidoscope indicated by the triangle drawn in the pattern above. If you place three mirrors around this kaleidoscope, you will see the entire pattern replicated from this triangle. The symmetry of this pattern is kaleidoscopic with signature

There are two 4's in the symbol because there are two different kinds of 4-fold kaleidoscopic points. The 2 in the symbol refers to the 2-fold kaleidoscopic point.

The fact that there can be several different kinds of kaleidoscopic points of the same order forces us to make it clear what *same kind* means for such points.

We say that any two features of a pattern are of the same kind only if they are related by a symmetry of the whole pattern. In other words, features of the same kind are those that *appear* the same to us, because some rigid transformation shifts one feature to the other yet leaves the pattern as a whole unchanged.

On the other hand, we can distinguish between the two 4-fold kaleidoscopic points in this kaleidoscope — they obviously appear different in the pattern, and moving one to the other shifts the pattern as a whole.

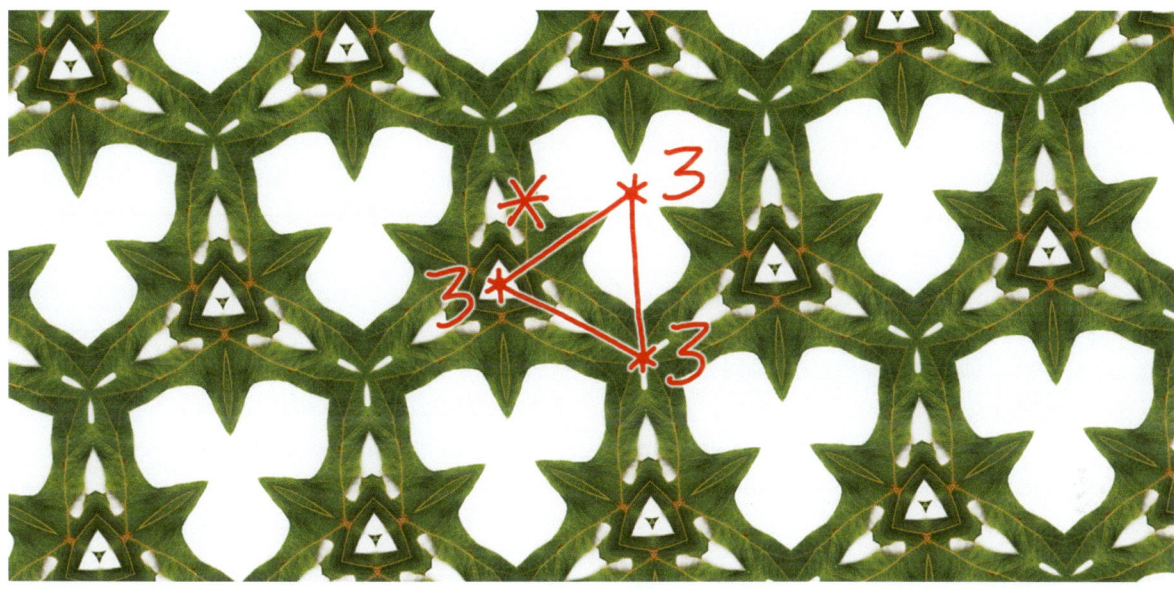

You can practice finding kaleidoscopes on these patterns by looking for the smallest region that can be enclosed by mirror lines.

When we draw all the mirror lines in the pattern above, we find that its kaleidoscope is a triangle with three different kinds of 3-fold kaleidoscopic points on its corners. Its signature is ✱333.

When we draw all the mirror lines in the pattern below, we find that its kaleidoscope has four different kinds of 2-fold kaleidoscopic points and its signature is ✱2222.

Gyrations

To find the kaleidoscopic part of the signature, we mark all of the mirror lines and locate the smallest kind of region that they enclose — a kaleidoscope. We then read out the different kinds of kaleidoscopic points around a kaleidoscope. When you mark the mirror lines in the pattern below you will find a kaleidoscope like the one we have drawn.

Because all the kaleidoscopic points in the pattern are of the same kind — any two of them are related by a symmetry of the whole pattern — we discover that the kaleidoscope is only of type *3 rather than *333.

However, the symmetries of this pattern are not purely kaleidoscopic! There is a new feature — a 3-fold rotational symmetry.

Let's look at this more closely. The pattern would be unchanged if the whole plane were to be rotated through 120 degrees around the gyration point marked 3 in the middle of the kaleidoscope. The same is also true of the kaleidoscopic point marked 3, but we've already accounted for this by calling it a 3-fold kaleidoscopic point — this rotation is "done by mirrors." Since the pattern has one kind of 3-fold gyration point and a kaleidoscope with one kind of 3-fold kaleidoscopic point, its signature is

3*3

You can turn to page 138 to see that the orbifold for this pattern has a boundary * with a marked point 3 upon it, and a special point 3 in its interior.

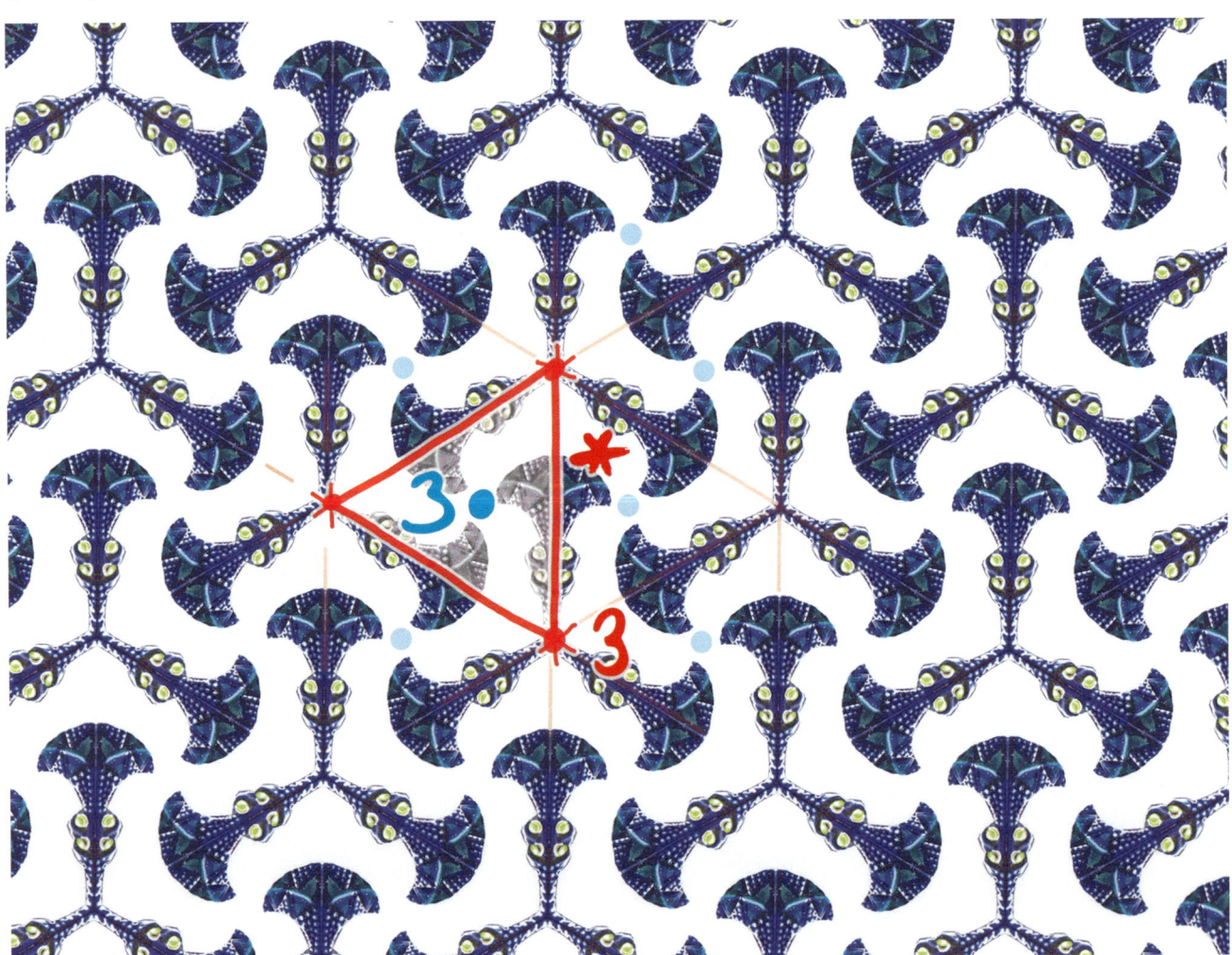

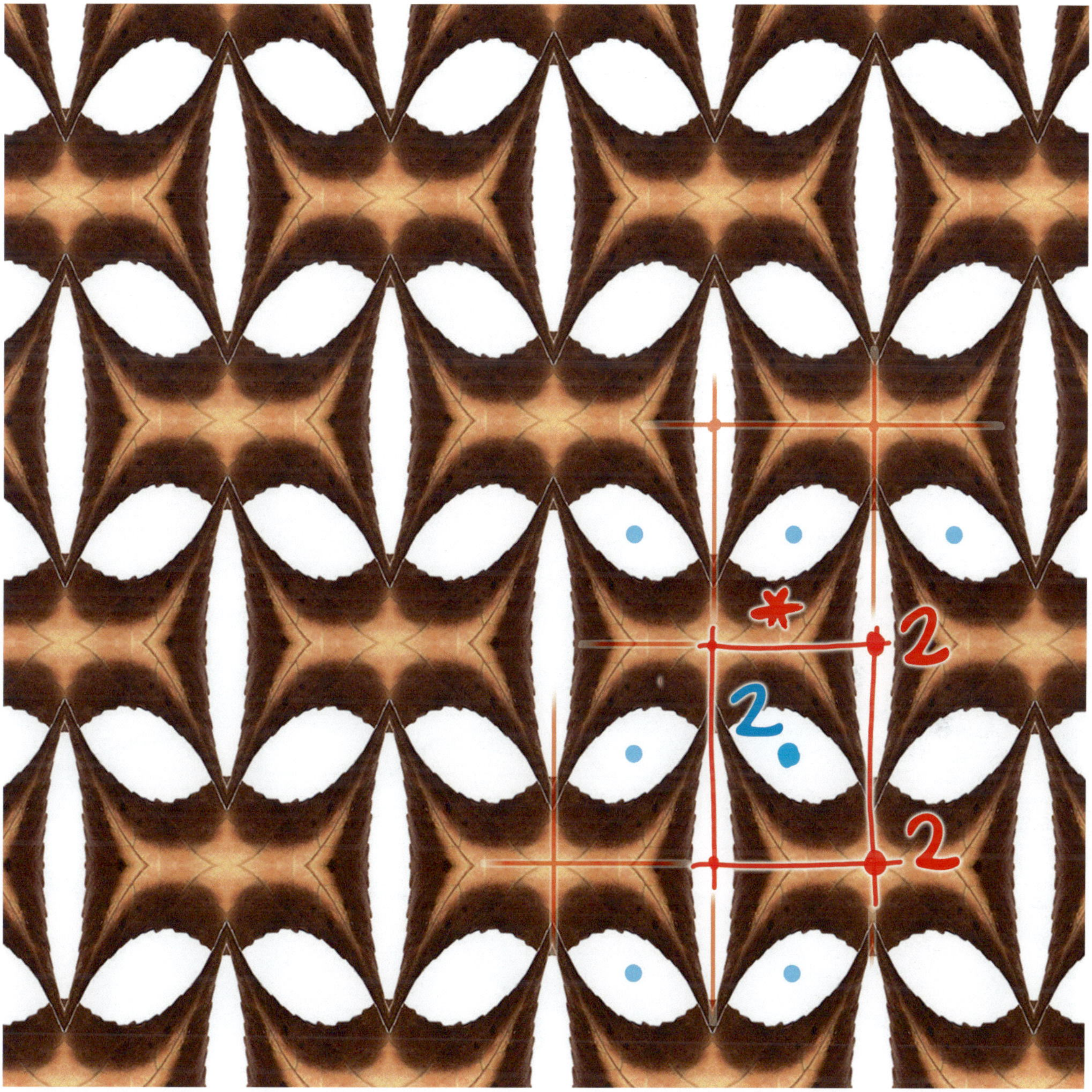

When we mark the mirror lines in this pattern, we find a kaleidoscope like the one we've drawn. There are 2-fold kaleidoscopic points around this kaleidoscope, but there are only two different kinds. Together the kaleidoscope and its two kaleidoscopic points are designated *22. In the center of the kaleidoscope, there is a 2-fold gyration point which we mark as 2. All together, this pattern has signature

Once you are familiar with this notation, you can tell immediately that a pattern with one kind of 4-fold gyration point and one kind of 2-fold kaleidoscopic point, like the pattern on this page, has signature

4*2

You can cover up our drawing on the pattern and practice looking for the signature yourself. Draw in enough mirror lines to find a kaleidoscope, and identify the kaleidoscopic points and gyration points in this pattern. Check that there really is only one kind of each.

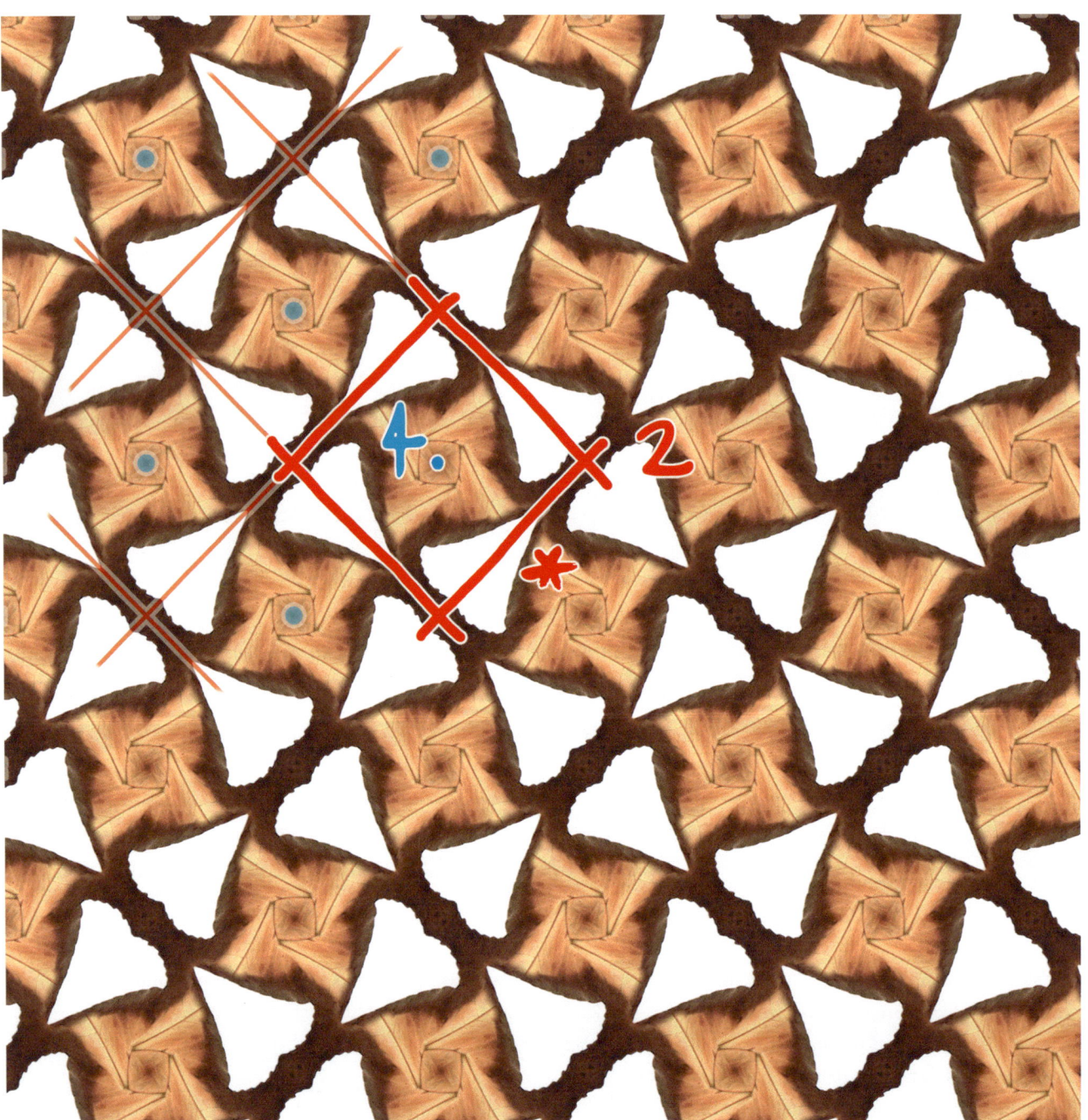

On this page we see a pattern that has no mirror lines and no kaleidoscopes, but has many gyration points. You can check that all of the gyration points are 3-fold and that there are three different kinds. The pattern has signature

333

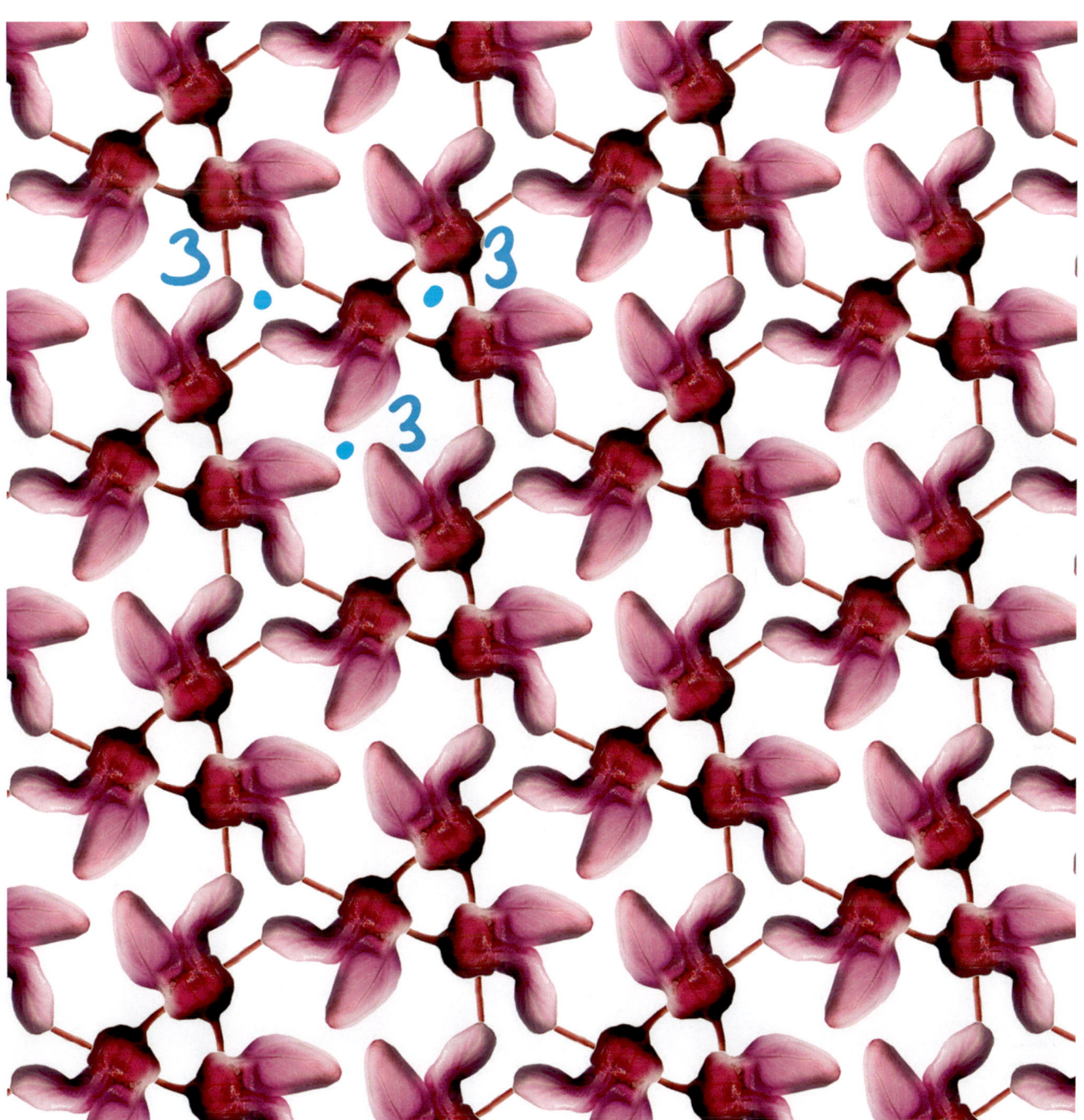

More Mirrors and Miracles

The two kinds of features that we have discussed, kaleidoscopes and gyration points, are easy to spot with a little practice — you can try your hand on the patterns at the end of this chapter. On the next few pages we'll meet a few extraordinary features that are harder to spot, because they really describe the building blocks of a pattern's orbifold. In the next chapter, the Magic Theorem will help us find them.

All the kaleidoscopes that we've seen so far have been defined by polygons enclosing part of our pattern, but that's not the only type there is. A single kind of mirror line that has no other mirror lines crossing it is a kaleidoscope with signature *.

You can check that this pattern has two different kinds of mirror lines — it has two different kaleidoscopes in it. Its signature is

We're also seeing something else for the first time here. In this pattern, the smallest subregion marked off by mirror lines is an infinite strip! There are several new features to be found in patterns like this one, which will be presented in this section and the next.

In the later chapters of this book we will learn that each kaleidoscope * records a boundary of the orbifold of a pattern. If you turn to Chapter 9 now, you can see that the signature of this pattern ** describes the two boundaries of its orbifold.

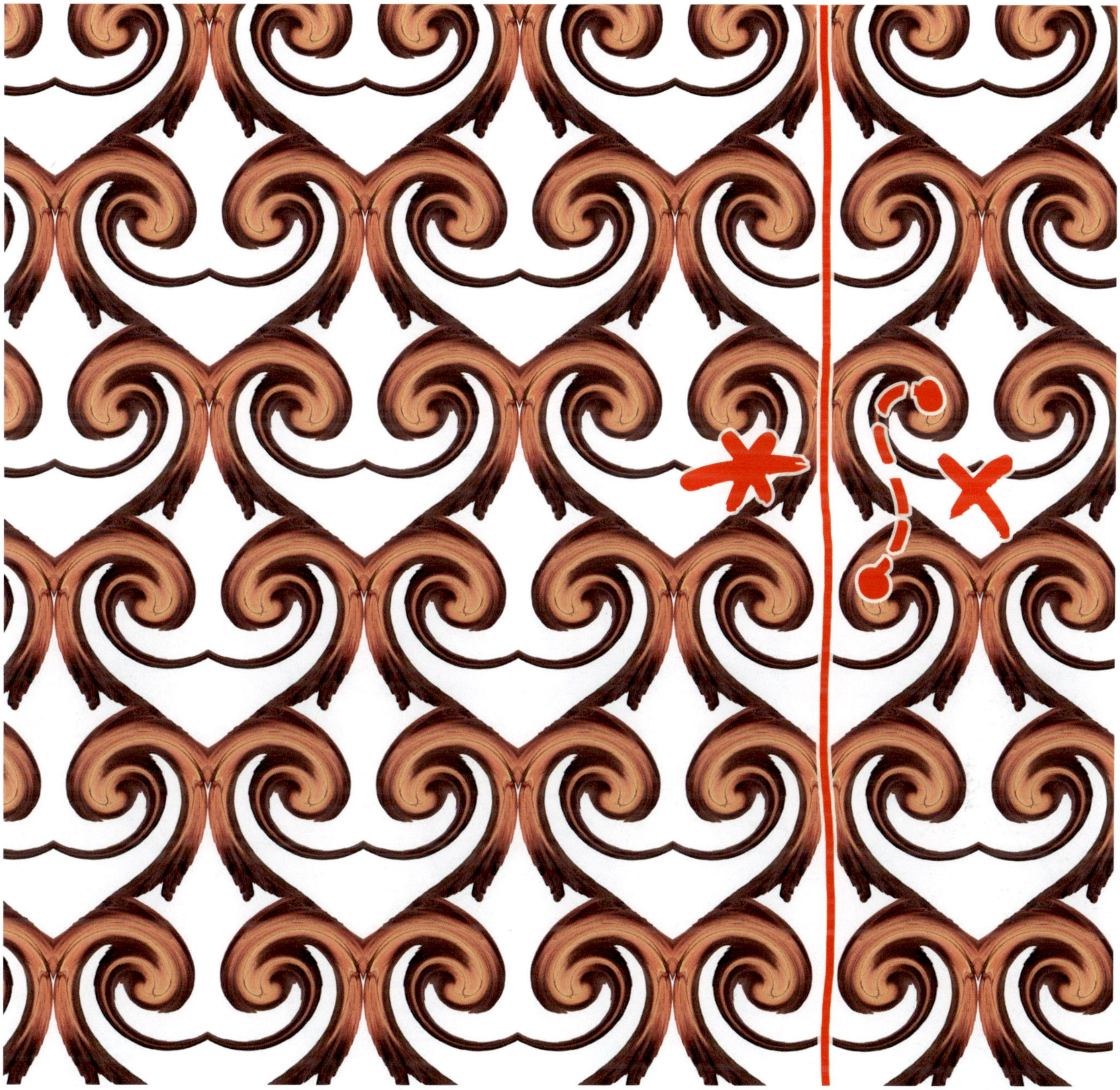

At first, this pattern looks very much like the one on the facing page. None of its mirror lines intersect, and the smallest subregion bounded by mirror lines is again infinite. But in this figure there is only one kind of mirror line!

And, there's a *miracle* here! There is a path from a left-handed spiral to a right-handed spiral that does not go through a mirror line. We will record the presence of such a path by a red cross (×) in the signature. We call this a "mir-

rorless crossing" or, for short, a *miracle*, and indicate it in figures by a red dotted line and cross.

This pattern has both mirrors and miracles, but only one kind of each, so its signature is

Turn to page 141 to see that the orbifold of this pattern is a Möbius band!

Just as we can have two different kinds of mirror we can have two miracles, as in this pattern which has signature

There are more than two paths from left-handed to right-handed spirals, but all of them can be made up of combinations of identical copies of the ones we've marked. Amazingly, the orbifold surface for this pattern is a Klein bottle (pages 127 and 143), though we must wait until Chapter 8 to learn how that surface is composed of two ×'s.

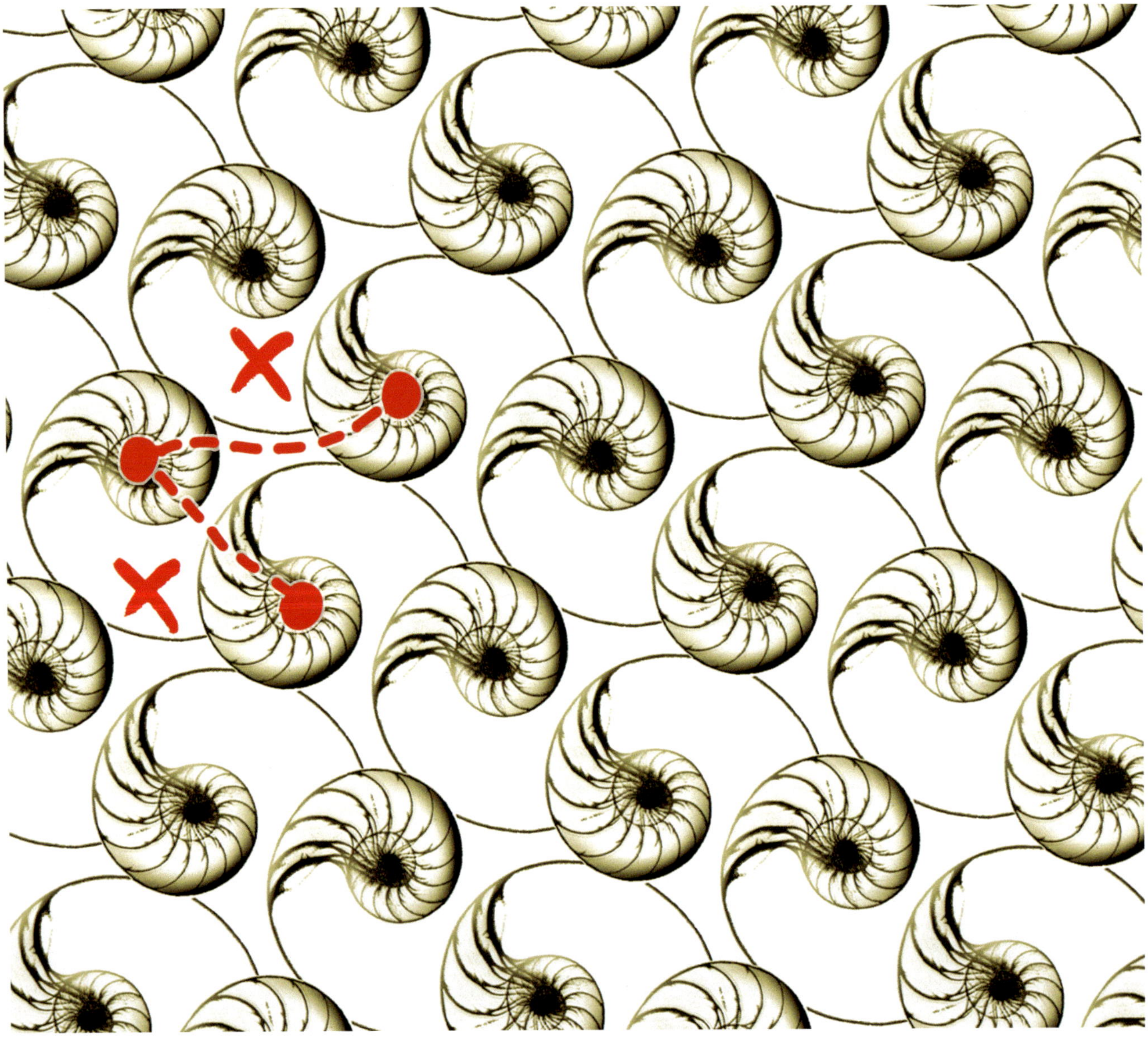

Wanderings and Wonder-Rings

Just as a miracle is a repetition-with-reflection of a fundamental region that's not "explained by" mirrors, it's possible to have a fundamental region repeated without reflection in a way that's not explained by gyrations, mirrors, or miracles, but by a simple shift in some direction.

In fact, such repetitions always come in pairs, in two different shift-directions, like the pair drawn on the pattern on this page. We call a pair of this kind a "wonderful wandering".

When we draw more copies of these pairs we can see they form a "wonder-ring," which we denote

On page 143 we roll these patterns up to see that their orbifolds really are ring-shaped — they are toruses!

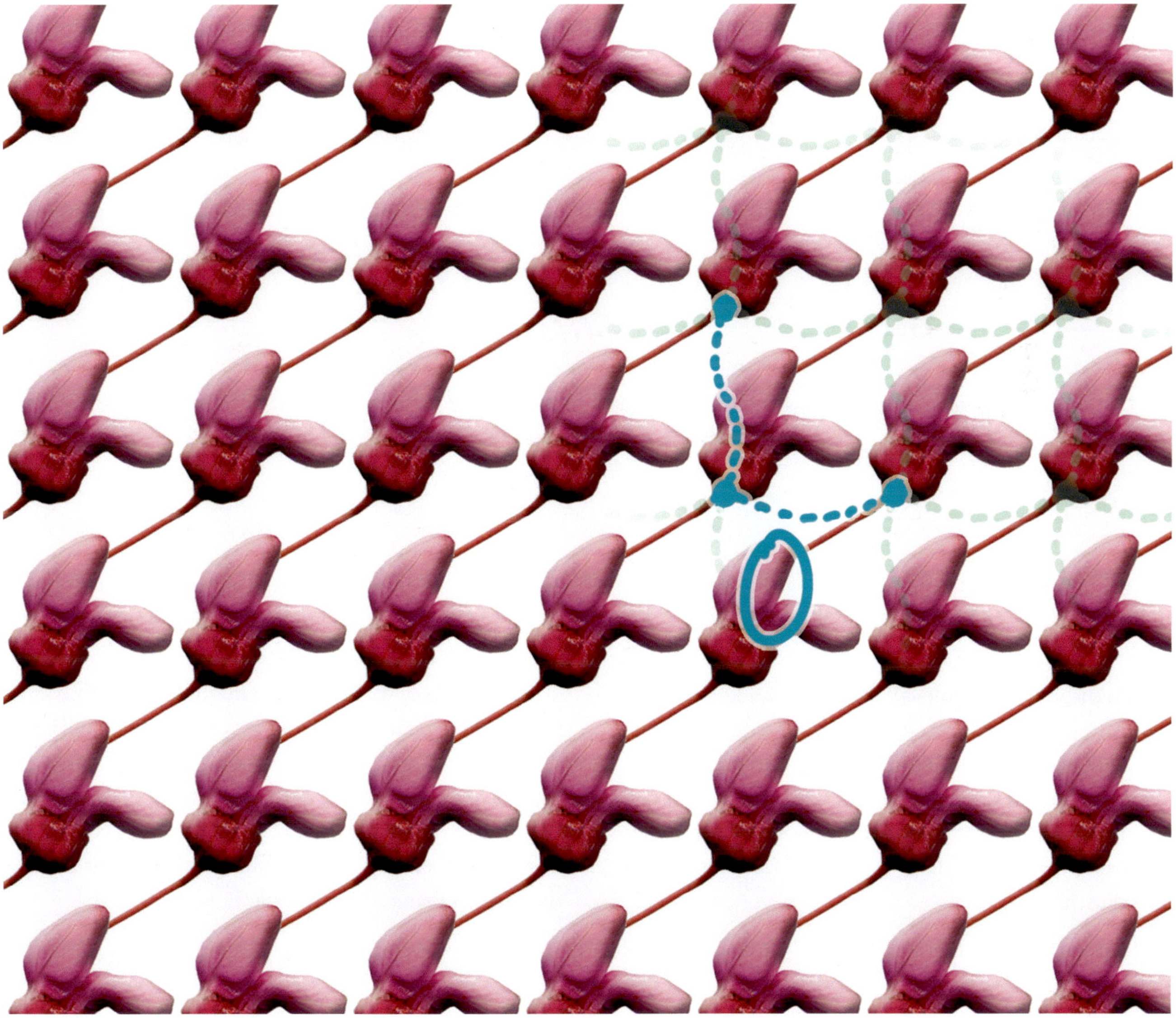

The Four Fundamental Features!

It is a remarkable fact that wonders, gyrations, kaleidoscopes, and mirrors suffice to describe all the symmetries of any pattern whatsoever, as we shall show in the next chapter. We therefore call them the *four fundamental features*. You obtain the signature of a pattern just by writing down whichever of these features it has.

Up to this point, we've used **blue** for wonders and gyrations, since these preserve the t**ru**e orientation of a fundamental region, and red for kaleidoscopes and miracles, since these **re**flect. However, you can write these in black ink if you always write them in the same order, since then you'll be able to work out which colors they should be.

The table below lists the four fundamental features in the appropriate order and the codes we use to represent them in the signature. In the later chapters of this book, we will see that these features record the building blocks of a pattern's orbifold.

Spherical and planar patterns use these features sparingly, but nearly any orbifold symbol describes *some* symmetry type if we include patterns in the hyperbolic plane as well (Chapter 10).

The Features of a Pattern

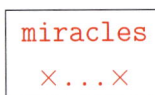

Where Are We?

In this chapter, we have described the four features of repeating plane patterns and introduced the *signature* that describes which of them appear in a given pattern. In the next chapter, we introduce a Magic Theorem which determines what combinations of features are possible for the signatures of plane patterns. There will be plenty of examples for you to practice on yourself.

In Chapter 6, we will see that our notation really records features of a pattern's orbifold and will better understand their nature. Using tools from topology, we can work out what features are possible for the orbifolds by a simple accounting of their "cost," the subject of the next chapter and its Magic Theorem. Chapter 9 shows orbifolds for many of the patterns that are in this book.

A Quiz

The features of these patterns are gyrations (with a signature like **632**), a kaleidoscope (like *333), or a mix of the two (like 4*2).

You can find the signatures of patterns like these in two steps. With a red pen, sketch in mirror lines and find a kaleidoscope, if there is one. As soon as you've found a region bounded by mirror lines, you can restrict your search to that region. Note down the orders of the different corners of the kaleidoscope — and don't double-count corners of the same kind! Next, with a blue pen mark any gyration points, where the pattern has rotational symmetry, but no mirror line. Note their order, and list out all of the features you have found. Congratulations: you have found the signature of your pattern! (Answers appear at the top of page 29.)

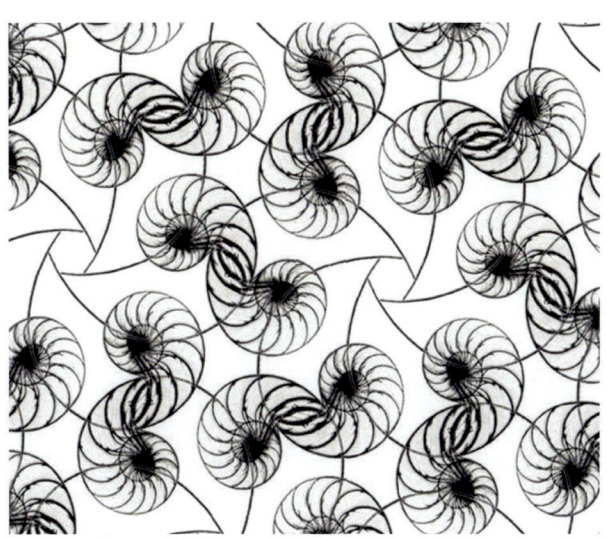

More patterns to analyze!

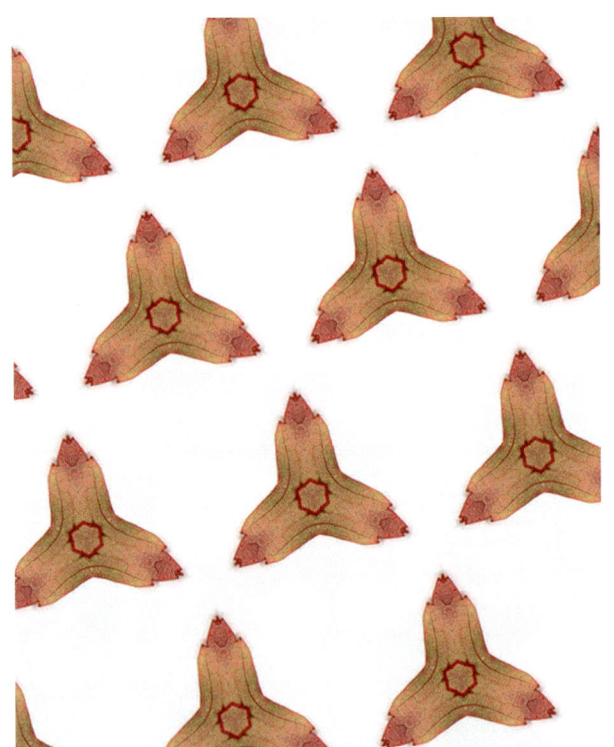

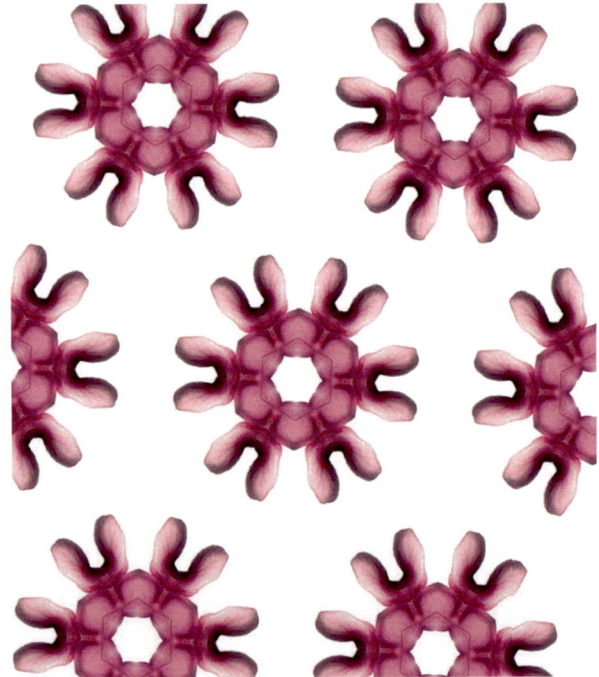

Here are our signatures for these patterns. Top row: **2∗22**, **333**, **∗333**, and **4∗2**. Second row: **4∗2**, **632**, **∗632**, and **2222**.

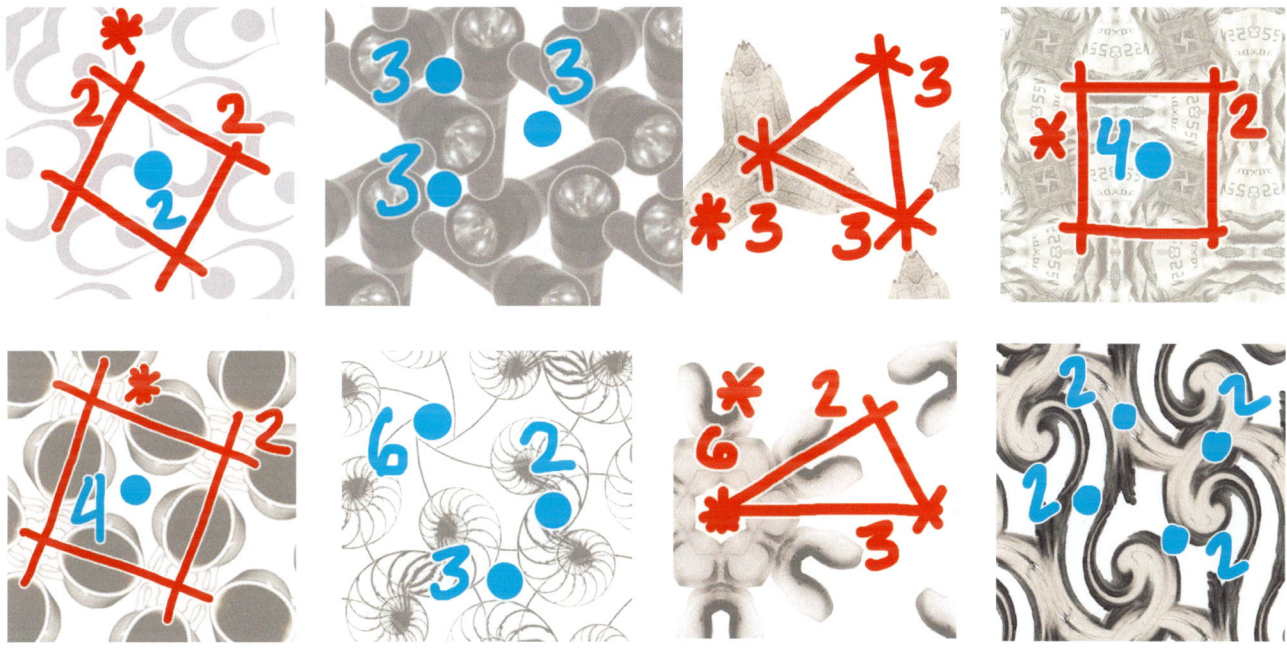

Extraordinary symmetry types:

The patterns in the quiz were more "ordinary": Their features are all gyration points and kaleidoscopes. None of these patterns has an "extraordinary" feature, like a miracle (×) or a wondering (○), or more than one kaleidoscope.

Analyzing patterns with features like these takes more care. The signatures are recording topological features of the pattern's orbifold, and some of these cannot be found just by looking at one location in the pattern. In the next chapter, we will use the Magic Theorem to help us find these "extraordinary" signatures. Meanwhile, here are three patterns with extraordinary symmetry type for you to practice on. The signatures for these patterns are on page 57.

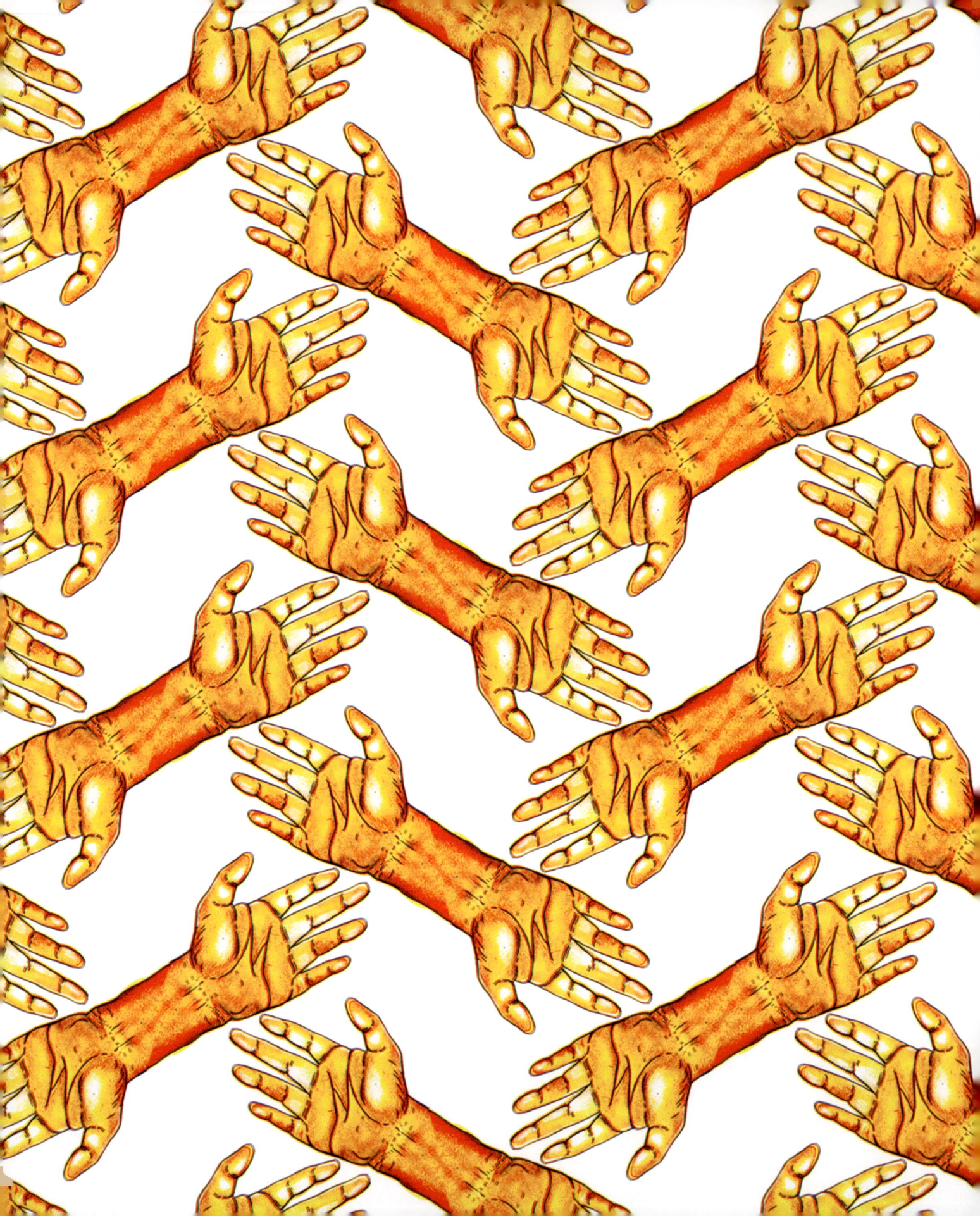

Chapter 3
The Magic Theorem

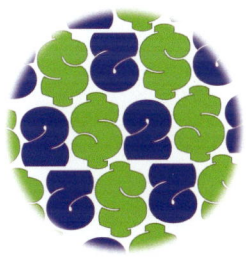

In the last chapter we introduced the four fundamental features of symmetry types for repeating patterns. From now on we shall often specify the symmetries of a pattern just by giving its signature (which lists its features). We haven't yet said *why* just these particular features are so fundamental — and we won't, until Chapter 8 — nor have we found just which signatures arise.

In this chapter we'll introduce you to the "Magic Theorem", use it to show that just 17 signatures are possible for plane repeating patterns, and then deduce that such patterns come in just 17 types. The proof of the Magic Theorem itself is something else you'll have to wait for!

Everything Has Its Cost!

It turns out to be a good idea to associate a cost to every symbol in the signature, as shown in the table on the right.

Why is this? Because, as we shall see in the next few chapters, there are Magic Theorems that tell us what signatures are possible in terms of their costs.

Symbol	Cost ($^\$$)	Symbol	Cost ($^\$$)
$\circ$	2	$*$ or $\times$	1
2	$\frac{1}{2}$	2	$\frac{1}{4}$
3	$\frac{2}{3}$	3	$\frac{1}{3}$
4	$\frac{3}{4}$	4	$\frac{3}{8}$
5	$\frac{4}{5}$	5	$\frac{2}{5}$
6	$\frac{5}{6}$	6	$\frac{5}{12}$
$\vdots$	$\vdots$	$\vdots$	$\vdots$
N	$\frac{N-1}{N}$	N	$\frac{N-1}{2N}$
∞	1	∞	$\frac{1}{2}$

Costs of symbols in signatures.

(opposite page) The Magic Theorem not only classifies signatures, but helps us determine the signature of a pattern. The signature **22×** of this pattern, like that of all planar patterns, costs exactly $^\$$2.

With the costs from the table on page 31, here is the Magic Theorem we'll use in this chapter.

Theorem 3.1 (The Magic Theorem for plane repeating patterns)
The signatures of plane repeating patterns are precisely those with total cost $^\$2$.

For example, the first pattern we analyzed on page 13 had signature $*632$, which has cost

$$^\$1 + \frac{5}{12} + \frac{1}{3} + \frac{1}{4} = {}^\$2.$$

We saw this next pattern on page 18. It has signature $3*3$, which costs

$$^\$\frac{2}{3} + 1 + \frac{1}{3} = {}^\$2.$$

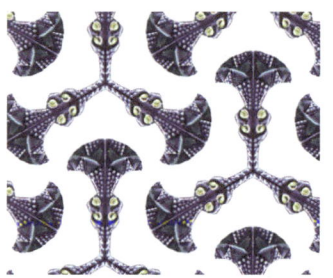

The third pattern, from page 19 has a kaleidoscope with two different 2-fold kaleidoscopic points and a 2-fold gyration point. Its signature is $2*22$, with cost

$$^\$\frac{1}{2} + 1 + \frac{1}{4} + \frac{1}{4} = {}^\$2.$$

Finally, the signature of the last pattern is $\times*$, with cost

$$^\$1 + 1 = {}^\$2.$$

Later in this book we will see how tools from topology will help prove the Magic Theorem. In this chapter we will use the theorem to list all possible signatures for repeating planar patterns. We will find there are seventeen, listed on page 35.

The fact that the signature of a planar pattern always costs $^\$2$ can help us check that the signature we have found for a pattern is correct; it can also help to complete it! For example, all we can see at first is that there are two kinds of 2-fold gyration points in the pattern below. But, **22** would only cost $^\$\frac{1}{2} + \frac{1}{2} = {}^\$1$, so there should be an extra dollar's worth to be discovered. Indeed there is! The pattern below is the same as its mirror image although it has no mirror line, so there must be a miracle $\times$ instead! We look at this more closely in the figure at right: there's a symmetry that takes a leaf to a backwards copy of itself, and the path joining these is the required miracle, giving us the signature **22**$\times$ with a total cost of $^\$2$.

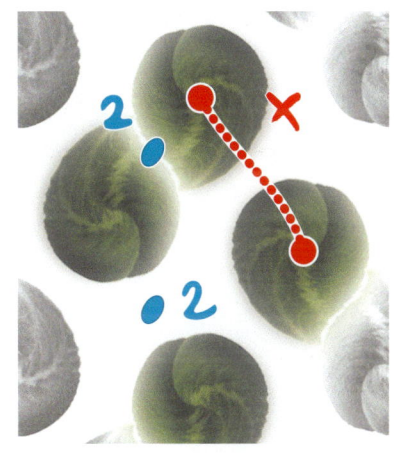

What is the signature of this pattern? Here there are also two kinds of 2-fold gyration points, which do not by themselves cost $^{\$}2$. The pattern is again the same as its mirror image, but a mirror, not a miracle, explains this, and the type is **22∗**, with cost $^{\$}\frac{1}{2} + \frac{1}{2} + 1 = {}^{\$}2$.

In this way the Magic Theorem can help us recognize the signatures of planar patterns, particularly those with miracles, wonder-rings, or more than one kaleidoscope.

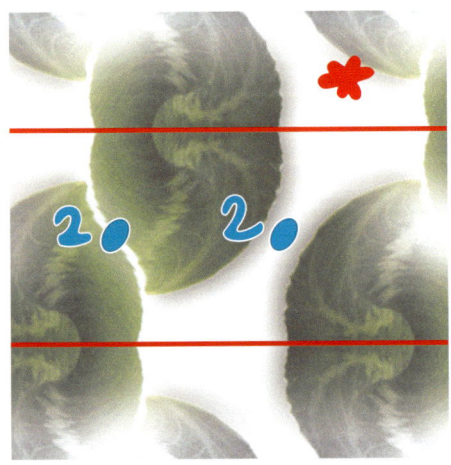

Finding the Signature of a Pattern

We can now exactly identify the signature of any repeating pattern on the plane by the steps listed at right. As we proceed, we write down the symbols in the signature. The Magic Theorem tells us that when we've found them all, the total cost of a planar pattern will be $^$2.

If we put our features in standard order, as listed in the table at the bottom of the page, we can tell at a glance which patterns have the same type.

If you encounter a tricky pattern, there are some things you should do to make your work easier. If two features are the same, you must only mark one of them. Sometimes it helps to label gyration points before labeling kaleidoscopes. Be sure there aren't any mirror lines inside the region bounded by a kaleidoscope, and don't forget that gyration points *never* lie on mirror lines!

The steps above work for any repeating pattern, on the sphere or in the Euclidean plane, or even in the hyperbolic plane. Here are some more hints that work just for patterns in the plane. There is one type of planar pattern with two kaleidoscopes and one with two miracles; if you're working with one of these, you should be able to see differences between these features by looking carefully at your pattern. You can use the fact that the total cost is $^$2 in several ways. You can stop when it reaches $^$2 (for instance, if you find a wonder), or if you have not yet reached $^$2, you will know that there must be more features to find.

1. If there are mirror lines, mark them in red, and examine the smallest regions into which they cut the plane. These regions are bounded by kaleidoscopes. Mark one of each type of kaleidoscope by a red $*$ and each kind of kaleidoscopic corner with its order. If you find a kaleidoscope, you can now restrict your search to any region bounded by mirrors.

2. Look for gyration points. In blue, mark just one gyration point of each type with a spot and its order.

3. Are there miracles? Can you walk from some point to a reflected image of itself without ever touching a mirror line? If so, a miracle has occurred. Mark such a path with a broken red line and a red cross $\times$ nearby.

4. Is there a wonder? If you've found none of the above in a planar pattern, then there is one: mark it with a blue wonder-ring $\circ$.

5. Finally, check that the total cost of the pattern's signature is $^$2, to see that all of the features are accounted for correctly.

In fact, as we shall soon see, all we have to do to find the signature of a pattern is *look at its orbifold surface!*

Just 17 Symmetry Types

Why are there just 17 types of symmetry for planar patterns?

We'll deduce this using only the Magic Theorem and some simple arithmetic. The calculations in the next few sections are very similar to those that answer the question, "How many different ways can I make change for a dollar if I use only quarters and dimes?" If the results at first seem mystical, try working through a few examples for yourself.

*632	*442	*333	*2222	**
			2*22	
				*×
	4*2	3*3	22*	
				××
			22×	
632	442	333	2222	o

The 17 symmetry types of planar patterns

The "True Blue" Types

If all symmetries of a pattern are obtainable by true motions, without flips, as in the patterns on these two pages, the signature will be entirely blue. If a blue string of digits **AB...C** is to cost $^$2, there must be more than two of them, since each costs less than $^$1. If there are exactly three, the values in the table of costs on page 31 show that the signature can only be one of **632**, **442**, or **333**. If there are more digits, the signature can only be **2222**, since each digit costs at least $^$\frac{1}{2}$.

Finally if there's a wonder-ring, the signature must be ○, since the ring already costs us $^$2.

The following figures illustrate the five true blue types:

632, **442**, **333**, **2222**, and ○.

If no digit is **2** then there must be three **3**'s and the type is **333**.

If the remaining two characters have *their* mean cost of $^\$\frac{3}{4}$, we get **442**.

If not, a second character must be **3**, and **632** is forced, since $^\$\frac{5}{6} + \frac{2}{3} + \frac{1}{2} = {}^\$2$.

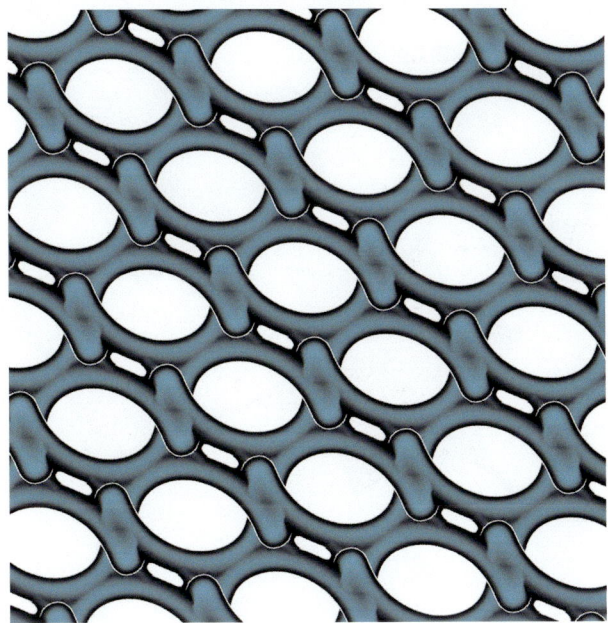

The only Euclidean type with four kinds of gyration points is **2222**, since $^\$\frac{1}{2} + \frac{1}{2} + \frac{1}{2} + \frac{1}{2}$ is already $^\$2$.

If there's a wonder ring ○ (costing $^\$2$), there can't be anything else.

The "Reflecting Red" Types

Now consider the signatures that are entirely red and have no crosses. They correspond to the previous cases because ∗AB...N costs $2 if and only if **AB...N** does:

$$^{\$}1 + \frac{A-1}{2A} + \ldots + \frac{N-1}{2N} = {}^{\$}2 \iff {}^{\$}\frac{A-1}{A} + \ldots + \frac{N-1}{N} = {}^{\$}2,$$

while there can only be one such signature (∗∗) with more than one star. This yields the five reflecting red types

∗632, ∗442, ∗333, ∗2222, and ∗∗.

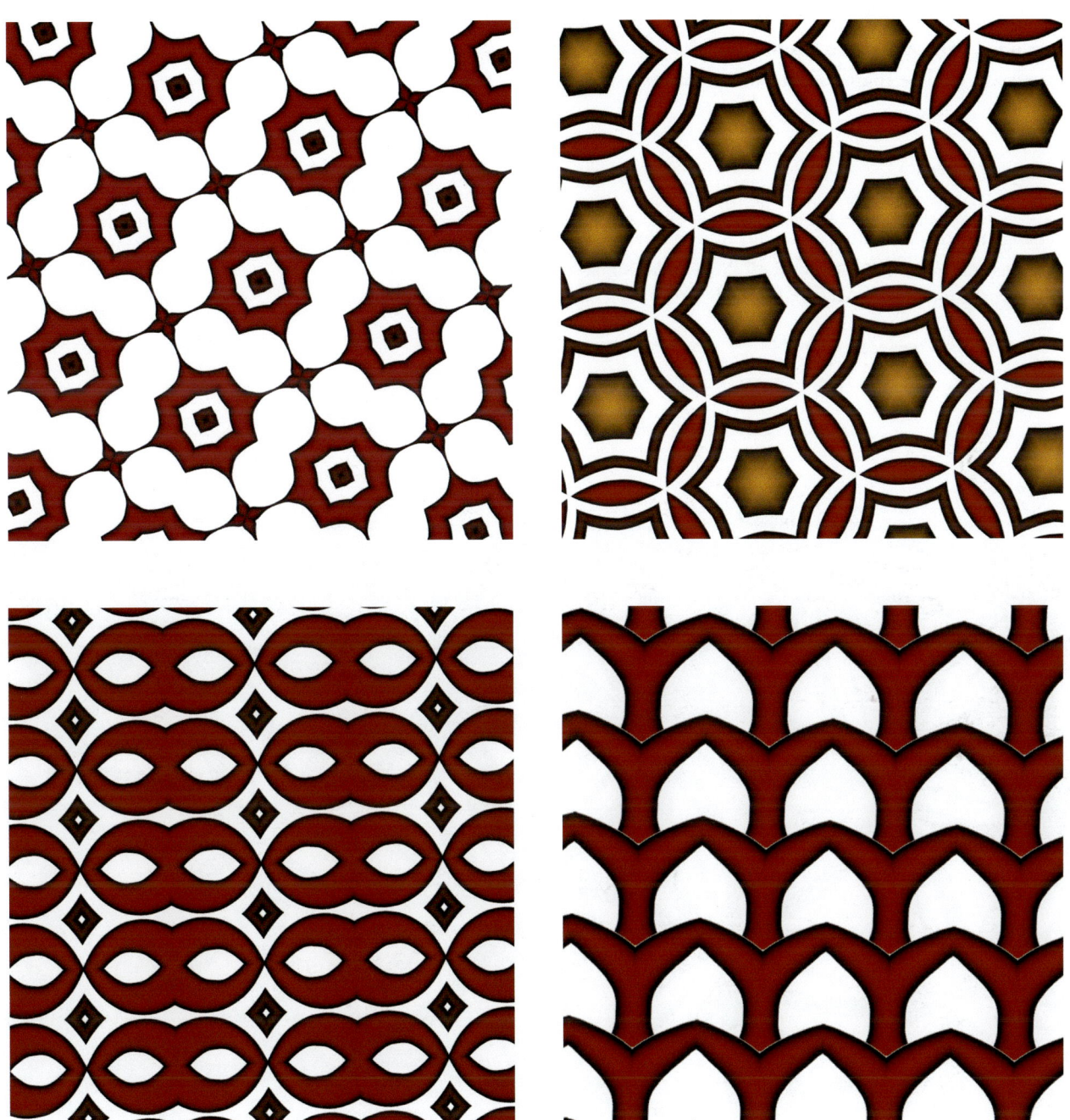

The all-red signatures, *333, *442, *632, *2222, and ** correspond exactly to the all-blue signatures 333, 442, 632, 2222, and ○, since each red digit costs half as much as the corresponding blue digit and a kaleidoscope (*) costs half of $2.

Exercise. Match the symmetry types to these patterns.

The "Hybrid" Types

The remaining signatures either mix blue and red or include
× symbols. To help us enumerate these "hybrid" types, we
note that the "demotions"

$$\text{replace } \mathbf{N}* \text{ by } *\text{NN}$$
$$\text{replace } \times \text{ by } *$$

don't change the cost and must eventually lead to one of the
five previous cases. So, we can recover all these mixed signatures by making the inverse "promotions"

$$\text{replace } *\text{NN} \text{ by } \mathbf{N}*$$
$$\text{replace a final } * \text{ by } \times$$

in all possible ways.

*632	*442	*333	*2222	**
↓	↓		↓	↓
4*2	3*3		2*22	*×
			↓	↓
			22*	××
			↓	
			22×	

The following seven figures represent the mixed types

$$3*\ \ 3,\ 4*2,\ 2*22,\ 22*,\ 22\times,\ *\times,\ \text{and}\ \times\times.$$

Exercise. Find the symmetry types of these patterns.

We conclude that there are just 17 possibilities for the signature costing $^\$2$, and so just 17 symmetry types for repeating patterns on the plane.

The Magic Theorem states that the signatures of plane repeating patterns are precisely those with total cost $^\$2$, and so implies that there are at most 17 symmetry types for a plane repeating pattern, traditionally called the 17 plane crystallographic groups.[†]

*632	*442	*333	*2222	**
			2*22	
				*×
	4*2	3*3	22*	
				××
			22×	
632	442	333	2222	o

How the Signature Determines the Symmetry Type

We have ignored some details. To what extent can we generate a pattern from its signature? This is a real problem, as we shall see in the spherical case[‡], but the answers in the plane case are easy. In the end, they depend only on the existence of rectangles and triangles with given angles, provided that those angles have the correct sum of π.

For instance, a pattern with signature *632 must be generated by reflections in the sides of a triangle with angles $\frac{\pi}{6}$, $\frac{\pi}{3}$, and $\frac{\pi}{2}$. All triangles that satisfy this condition will be the same up to size, so up to similarity there's just one possibility for the symmetries of a pattern with signature *632. The same holds for signatures *333 and *442.

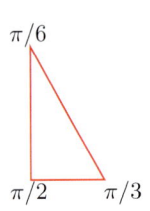

In the same vein, the symmetries of a pattern with signature *2222 are generated by the reflections in the sides of a quadrilateral whose four angles are $\frac{\pi}{2}$, that is to say, a rectangle. Here the set of symmetries is no longer unique up to scale; any one version can be continuously reshaped into any other by gradually varying this rectangle.

The result is that one set of symmetries can be continuously transformed into the other while consistently maintaining its type, not changing the signature. In technical language this kind of deformation is called an *isotopy*. Some planar symmetry types, like *2222 can be *isotopically reshaped*, distorting one pattern into another with the same signature, while others, such as *632 are rigid and cannot be reshaped.

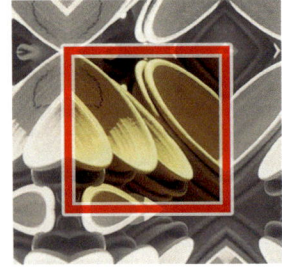

[†]In the table, the nonreflecting elements of any of these groups form its orientation preserving subgroup, at the bottom of the column.

[‡]In fact, with only the exceptions of **MN** and *MN with $M \neq N \geq 1$, *every* possible signature describes a symmetry type, but our Magic Theorems show that only a few are planar (cost $^\$2$) or spherical (cost < $^\$2$). All of the rest of the signatures (cost > $^\$2$) describe symmetry types in the hyperbolic plane (Chapter 10).

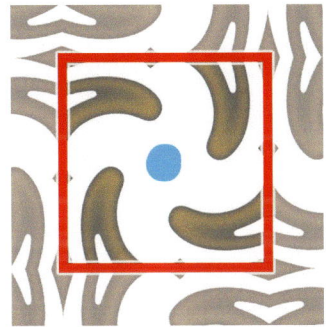

For 4∗2, four copies of a fundamental region combine to form a square bounded by a kaleidoscope of mirrors. Then reflections in the sides of that square generate the rest of the pattern. Up to rescaling there's really only one set of symmetries corresponding to 4∗2, and this type is rigid.

You can confirm for yourself that the argument given for 4∗2 is easily adapted to the rigid type 3∗3 and to the type 2∗22, which can be isotopically reshaped.

The signatures 632, 442, and 333 fix a lattice of gyration points in the same manner as their red counterparts fix a grid of mirror lines, and are also rigid. But both 2222 and ○ can be isotopically reshaped by a choice of parallelogram as a fundamental region.

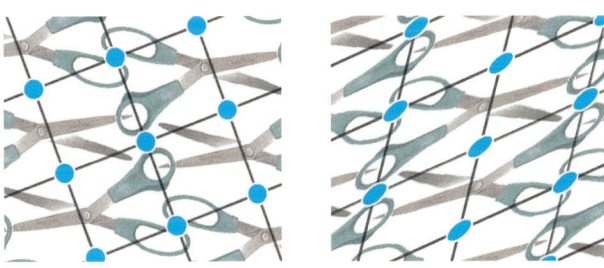

Case-by-case arguments like these work for all 17 types. Some experimentation will reveal that the shape of the remaining signatures is determined by choosing the shape of the rectangle to be used as a fundamental region.

Where Are We?

Using the Magic Theorem we've now shown that there are just 17 plane crystallographic groups.

As we said, you'll have to wait to see why the Magic Theorem is true.

The next two chapters will discuss the versions of it that apply to patterns on the sphere and to planar frieze patterns. Then we'll move on to the general theory, seeing that a pattern's signature is really recording the topology of the pattern's orbifold, and that the Magic Theorem describes restrictions on the topological features a planar pattern's orbifold may have.

Before we do that, we share some exercises and their answers. We include an outline of the steps here. Turn to page 35 for details and tips. On the next page we pause for physical kaleidoscopes.

To find the signature of a pattern,

- identify any kaleidoscopes — a chain or loop of mirrors — and the corner kaleidoscopic points upon them;

- identify any gyration points;

- identify miracles and wonder-rings;

- and for a planar pattern, check that the cost of your signature is $2.

Interlude: About Kaleidoscopes

Kaleidoscopes — the physical kind found in toy stores — were invented by Sir David Brewster in 1816. In a real kaleidoscope, displaying a properly repeating planar pattern at its end, its mirrors can only be arranged as shown at right. That is, the symmetry type's signature is just that of one of the reflecting red types *333, *442, *632, or *2222.

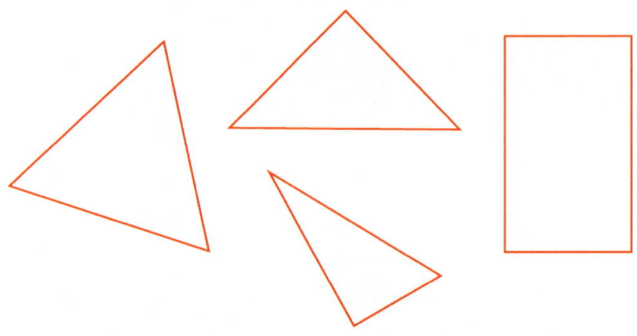

Here are some photographs taken inside of real kaleidoscopes. We hope you are inspired to obtain some mirrors and make a kaleidoscope yourself!

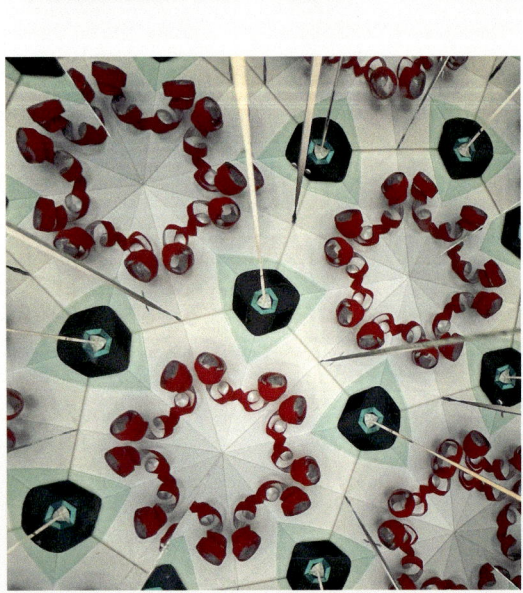

Exercises

1. We've told you how to find the signature of a pattern, but you'll want some practice to get it right! Let's start with these interesting patterns, which we saw on a chair seat, on a floor mat, in a hotel lobby, on a parking lot wall... Patterns like these are almost anywhere you look!

After you've worked out the signatures of these patterns for yourself, turn the page to check your answers.

Here are the signatures of the patterns on the previous page. Top row: **22***; **22x**;
and **2*22**. Second row: **442**; **22***; and ***442**. Third row: **2222**; **3*3**; **4*2**.

2. Repeating patterns on brick walls are always fun to analyze on a walk.
What are the signatures of these patterns?

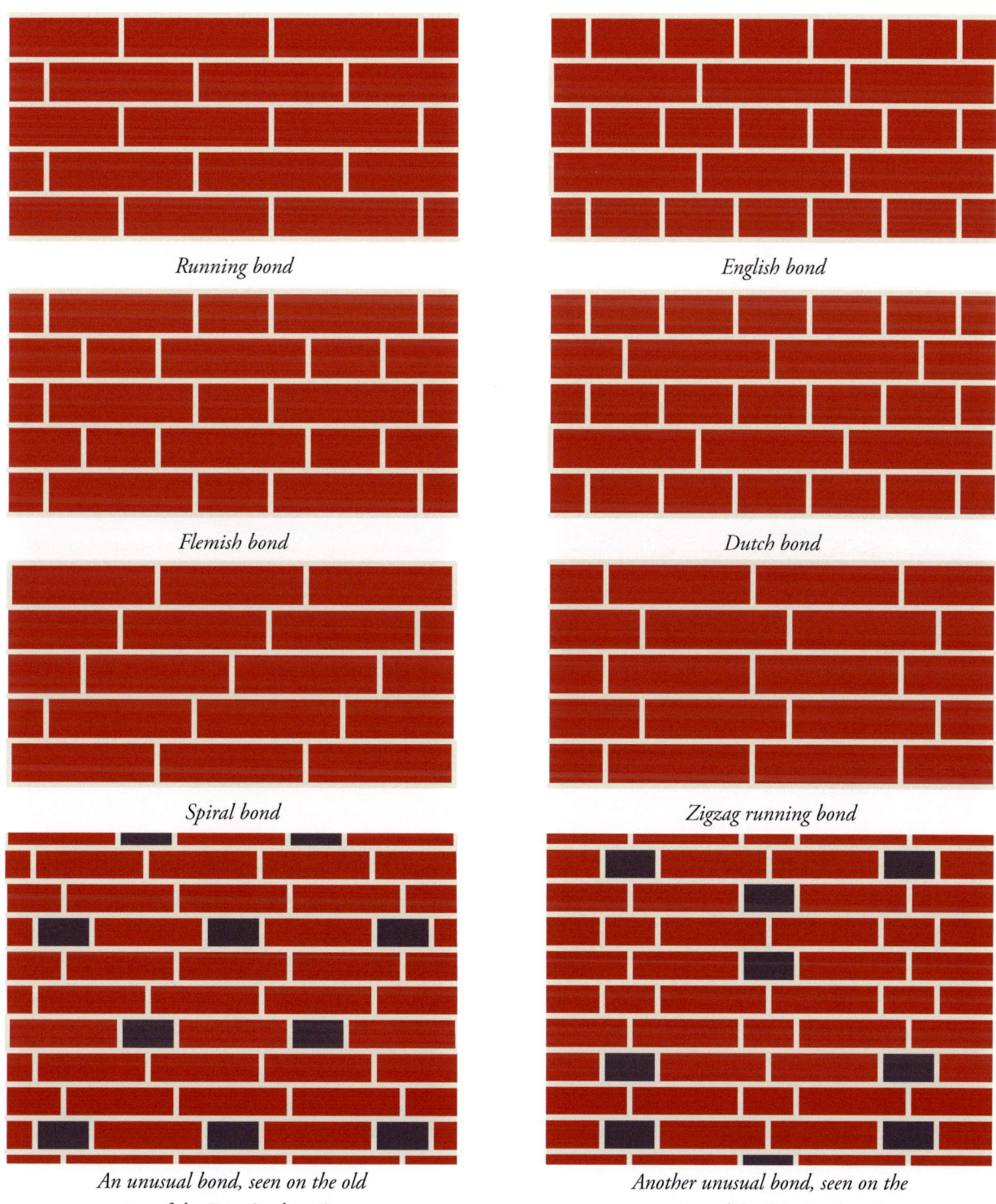

Running bond

English bond

Flemish bond

Dutch bond

Spiral bond

Zigzag running bond

*An unusual bond, seen on the old
section of the Frist Student Center*

*Another unusual bond, seen on the
new section of the Frist Student Center*

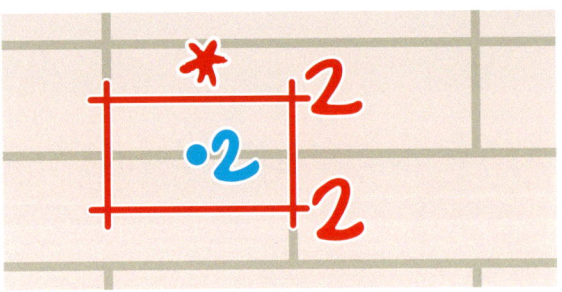

Running bond has type 2*2.

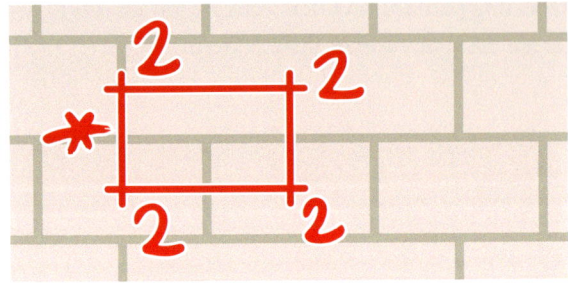

English bond has type *2222.

Flemish bond has type 2*2.

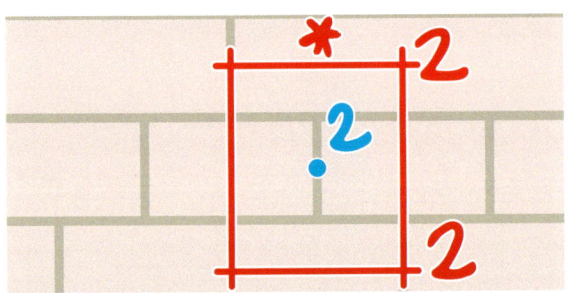

Dutch bond also has type 2*2.

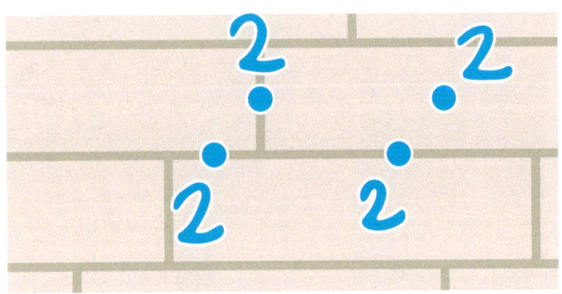

Spiral bond has type 2222.

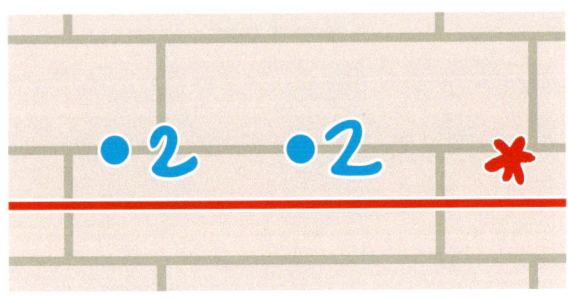

Zigzag running bond has type 22*.

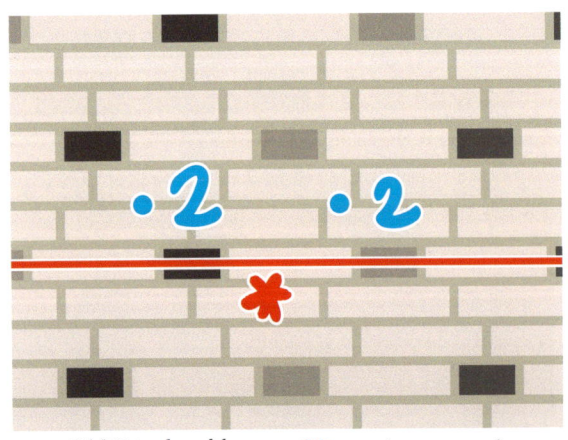

Old Frist bond has type 22* — its symmetries interchange bricks we've colored in the same way.

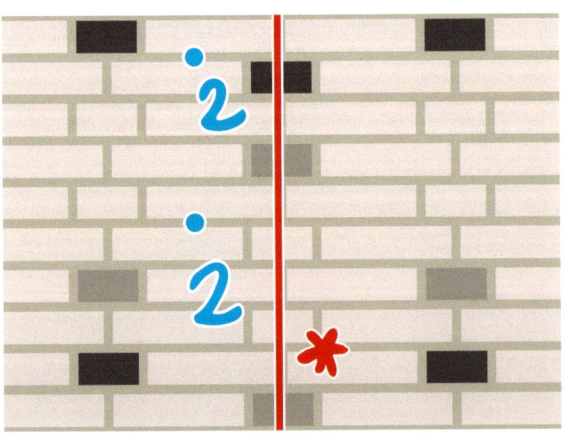

New Frist bond also has type 22*, with vertical mirror lines.

3. Find the signatures of these patterns! Turn the page to check your answers.

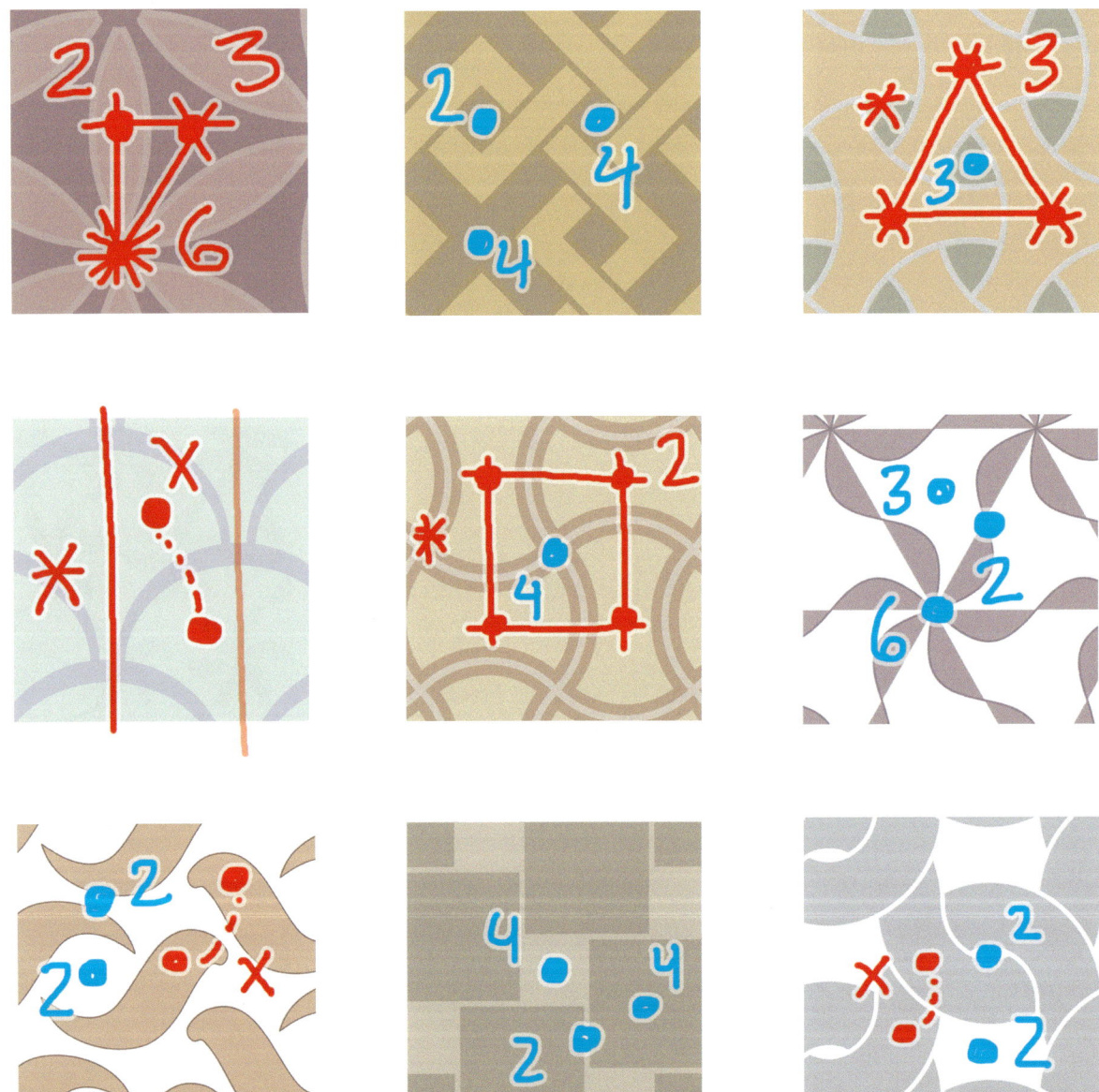

THE MAGIC THEOREM

Check your answers. Top row: *632, 442, 3*3. Middle row: *x, 4*2, 632.
Bottom row 22x, 442 and again 22x.

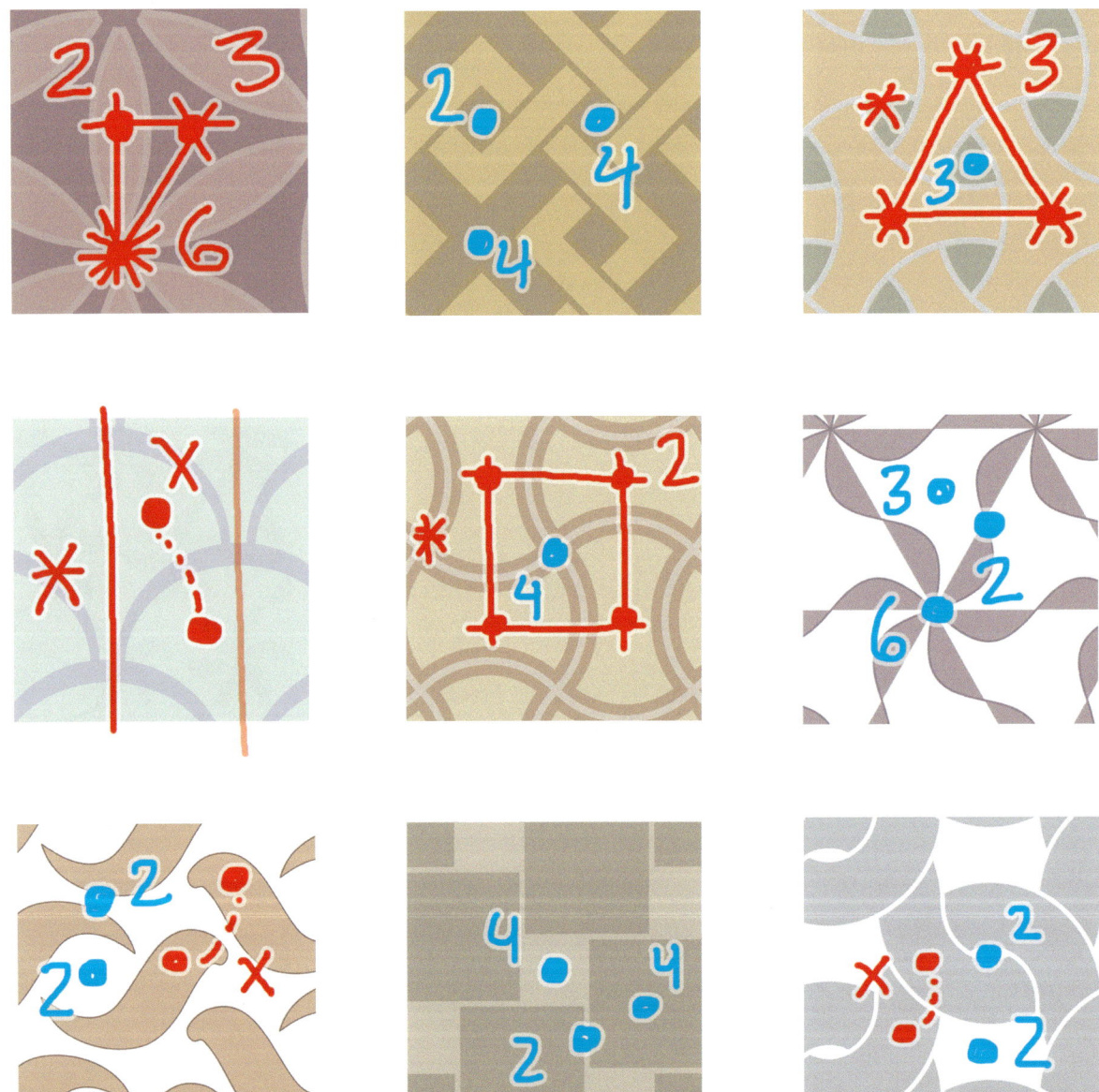

4. The placement of the dots changes the signatures of these patterns. Get out your markers and identify their signatures. Be careful not to get your eyes crossed!

Check your answers, and create your own dot patterns to analyze.

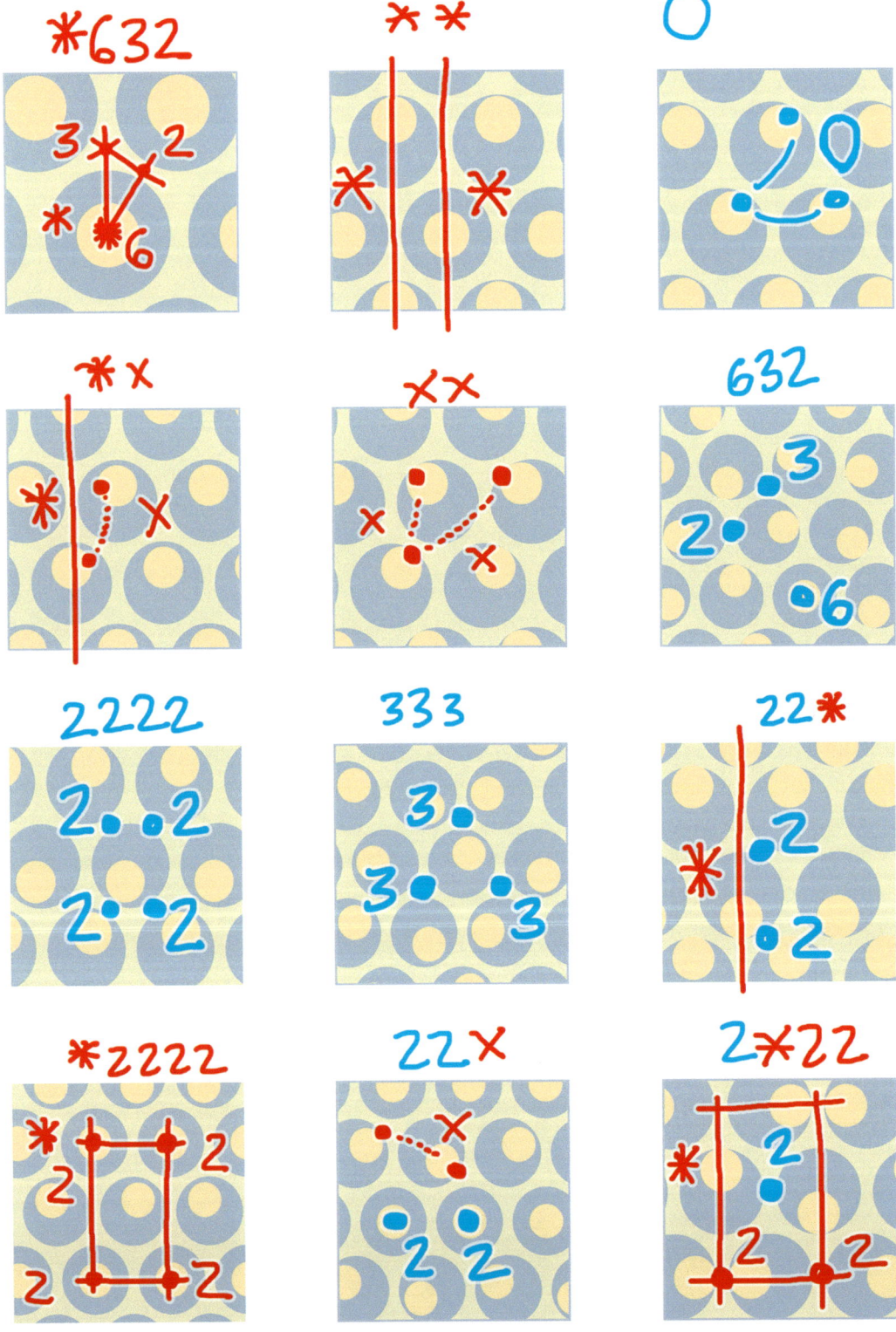

5. Here are more patterns that we found while walking about.
What are their signatures?

The signatures of the patterns on the previous page are: First row: ×∗; ○ but if symmetries may swap the darker two grays, ×∗; 22∗ unless we pay closer attention to the colors, in which case, ∗∗. Second row: ○; 333; 2∗22. Third row: 3∗3; 2∗22; and 2∗22.

6. Here are pavements, mats, and manhole covers that we've encountered — only a fraction of the beautiful geometric ornament that people have created around us.

How many different signatures can you find where you live?

The signatures of the patterns on the previous page: First row: 4*2; xx; 2*22.
Second row: 4*2; 2222; and 3*3. Third row: 2222; 442; 22*.

 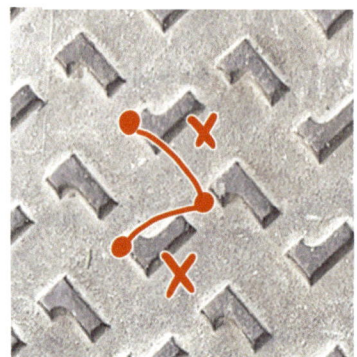 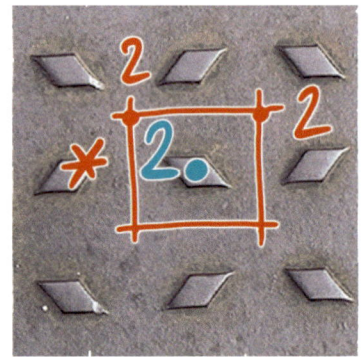

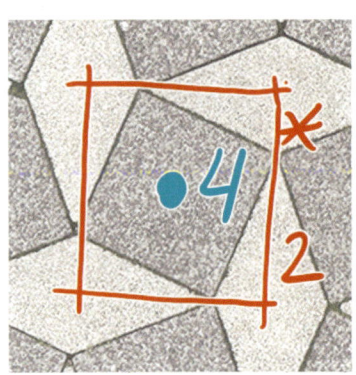 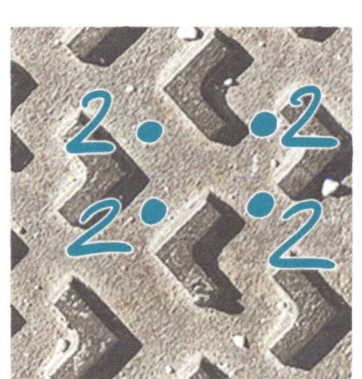 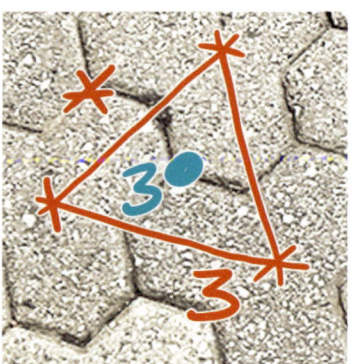

  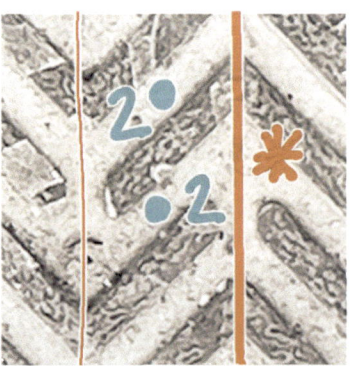

Here are the signatures for the regular planar patterns on page 8 of Chapter 1.
Top row **632**, **o**, and *632. Bottom row **22x** (that miracle can be hard to spot!),
333, and **442**.

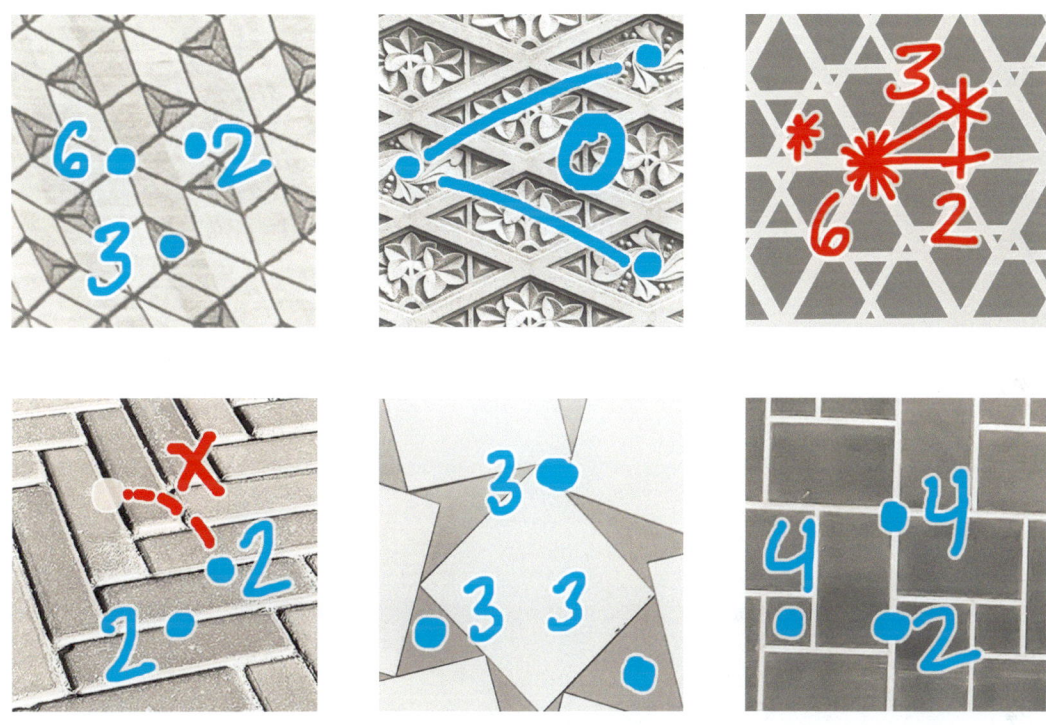

These are the signatures of the patterns at the end of Chapter 2, on page 29.

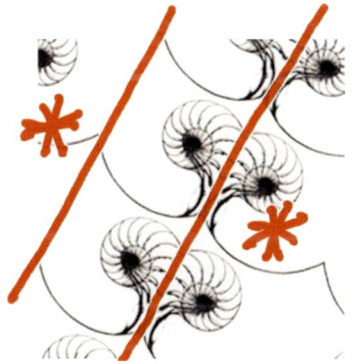

Chapter 4
Symmetries of Spherical Patterns

So far, we have discussed only symmetric patterns on planar surfaces. However, most of the symmetric things we encounter in our everyday lives aren't planar surfaces. Chairs, desks, boxes, and even people (roughly) are symmetric, but non-planar.

To find the features describing the symmetries of an object like a chair or table, we imagine it as resting inside a "celestial sphere" surrounding it. By studying spherical symmetries, we can understand the symmetries of everyday things.

For the chair (right) there is a single plane of reflection that intersects the sphere in a single mirror line — in other words, it has *bilateral symmetry*. The signature for the bilateral type of symmetry is ∗, because we see one mirror line on the surface of the sphere and it meets no other mirror lines.

We see from the table of costs on page 31 that this only costs $1, so it is cheaper than the plane crystallographic groups, which all cost $2.

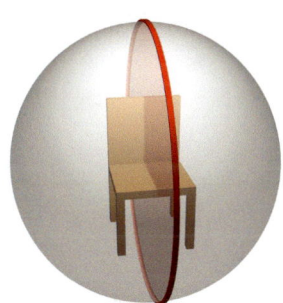

(opposite page) Four spherical patterns, with signatures ∗532, ∗432, and two with ∗2 2 11. The strange pair of eyeglasses (above) has signature x in its celestial sphere.

More complicated objects can have kaleidoscopic points, gyration points, and miracles. For the rectangular table at left, the mirror lines are two great circles that meet at right angles on the celestial sphere. On the sphere they have two intersection points. Both of these are 2-fold kaleidoscopic points. Therefore the symmetry type of this table has signature $*22$, costing

$$^\$1 + \frac{1}{4} + \frac{1}{4} = {^\$}\frac{3}{2},$$

again less than $^\$2$.

In just the same way, the celestial sphere of an ordinary box has three mirror lines, each pair of mirrors meeting at a right angle, as we show below at left. The mirrors form a kaleidoscope with three kaleidoscopic points of order 2 at its corners. The symmetry type of an ordinary box is $*222$.

It turns out that an important quantity is the change we get from $^\$2$, for which we will use the abbreviation ch. Thus,

$$ch(Q) = {^\$}2 - cost(Q).$$

For our chair

$$ch(*) = {^\$}2 - cost(*) = {^\$}2 - 1 = {^\$}1,$$

and the chair has $2 \div 1 = 2$ symmetries that do not change it: a reflection taking the right side to the left, and the act of doing nothing at all, which takes the left side of the chair to itself. We say this pattern has order 2, because it consists of two copies of its fundamental region.

Our table has change

$$ch(*22) = {^\$}2 - cost(*22) = {^\$}2 - 1 - 2\frac{1}{4} = {^\$}\frac{1}{2},$$

and the table has $2 \div \frac{1}{2} = 4$ symmetries and has order 4.

A box has change

$$ch(*222) = {^\$}2 - cost(*222) = {^\$}2 - 1 - 3\frac{1}{4} = {^\$}\frac{1}{4},$$

and we can count that a box has $8 = 2 \div \frac{1}{4}$ different symmetries. This type of spherical pattern has order 8.

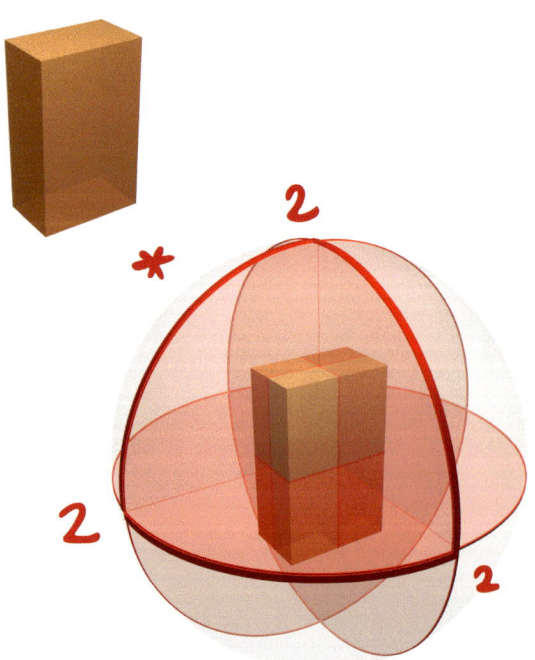

The signatures of the Euclidean planar patterns all cost exactly $2, so if you purchased any one of them with a $2 bill, you would get no change at all. But for spherical patterns, which have only finitely many symmetries, the rule is different: The change you get is precisely $2 divided by the number of symmetries.

Theorem 4.1 (The Magic Theorem for spherical patterns) *The signature of a spherical pattern costs exactly* $2 - \frac{2}{g}$, *where g is its order, or the number of its symmetries.*

In particular, the change is always positive, so the cost is always less than $2. We'll prove this in Chapter 6. In this chapter, we'll use it to derive the list of possible types of spherical pattern.

Our Euclidean Magic Theorem is really just a particular case of this, because there $g = \infty$, and so the change is $ch = \frac{\$2}{\infty}$, or 0. Thus, we don't really have two magic theorems but only one.

The 14 Varieties of Spherical Pattern

From the Magic Theorem we conclude that the spherical types are exactly

∗532	∗432	∗332	∗22N	∗MN
				N∗
		3∗2	2∗N	N×
532	432	332	22N	MN

Here M and N represent arbitrary positive integers, and there are infinitely many spherical symmetry types — but it turns out that there is a proviso: The types ∗MN and MN only happen when $M = N$. [†]

We allow M and N to be 1, with the convention that digits 1 can be omitted. This makes sense — a gyration point of order 1 or a kaleidoscopic point with exactly 1 mirror passing through it is uninteresting to us, so we let 1∗ = ∗11 = ∗. So that trivial symmetry 11 does not have an empty symbol, we denote it •. This is appropriate: In Chapter 1 we introduced

this • to indicate that a rosette symmetry fixes a point. The trivial symmetry fixes all of them.

As in Chapter 2, we proceed by first counting the all-blue spherical signatures, then the red ones, and finally those that involve both colors.

Opposite points are identical in this spherical pattern with signature ×=1×

[†]In fact, every other combination of features describes a two-dimensional symmetry type (Chapter 10). The Magic Theorems tell us if these are in the Euclidean plane, on the sphere, or in the hyperbolic plane.

The Five "True Blue" Types

Since the total cost of the signature must be less than $^\$2$, we cannot afford a wonder ring (○) or to have more than three digits (distinct from 1). The most general signature with fewer than three digits may be written **MN** by inserting **1**'s if necessary. Every such signature does cost less than $^\$2$, but according to the proviso it only corresponds to a symmetry type if **M = N**.

If there are exactly three digits, then one must be a **2**, or the cost is at least $^\$\frac{2}{3} + \frac{2}{3} + \frac{2}{3} = {}^\$2$.

If there are two or more **2**'s, the symbol is **22N** for some N, costing $^\$\frac{1}{2} + \frac{1}{2} + \frac{N-1}{N}$.

If there is just one **2**, then some other digit must be **3** since $^\$\frac{1}{2} + \frac{3}{4} + \frac{3}{4} = {}^\$2$. The remaining digit must be 3, 4, or 5 since $^\$\frac{1}{2} + \frac{2}{3} + \frac{5}{6} = {}^\$2$ and the signature of the pattern is **332**, **432** or **532**.

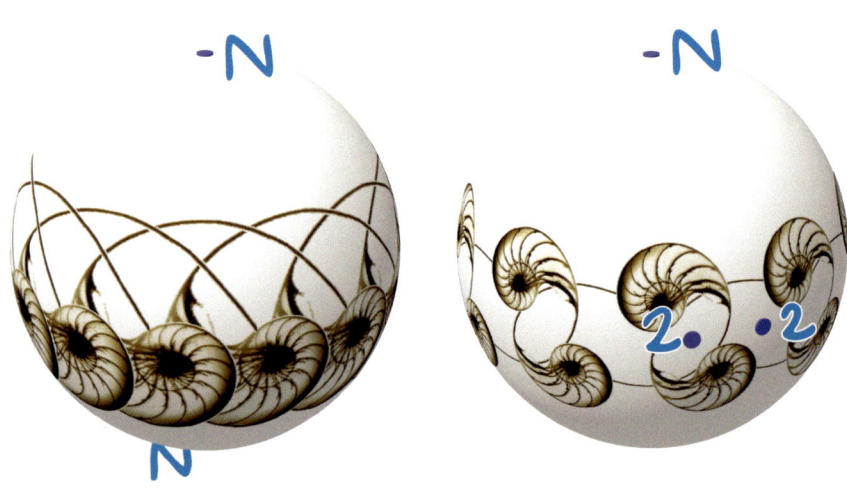

Signature **NN**   *Signature* **22N**

Signature **332**   *Signature* **432**   *Signature* **532**

The Five "Reflecting Red" Types

The all-red signatures for sphere patterns must have the form
∗AB...N since we can no longer afford two ∗'s. The ones
for which ch is positive are in perfect correspondence with
the true blue types, since $ch(∗AB...N)$ is exactly half of

$ch(AB...N)$, as we see from the following:

$$ch(∗AB...N) = {}^\$2 - 1 - \left(\frac{A-1}{2A} + \cdots + \frac{N-1}{2N} \right),$$

$$ch(AB...N) = {}^\$2 - \left(\frac{A-1}{A} + \cdots + \frac{N-1}{N} \right).$$

But remember the proviso: ∗MN exists only if $M = N$.

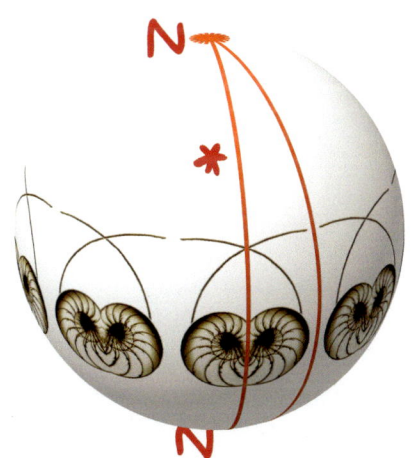

Signature ∗NN

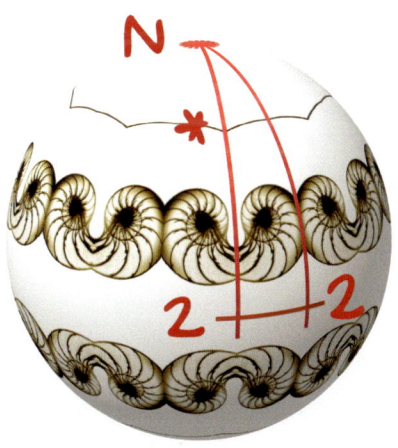

Signature ∗22N

Signature ∗332

Signature ∗432

Signature ∗532

The Four Hybrid Types

As in the plane case, the hybrid, mixed types must all be obtainable by promotion from the red reflective cases. Here are
all the possibilities:

*532	*432	*332	*22N	*NN
		↓	↓	↓
		3*2	2*N	N*
				↓
				N×

*Signature 3*2*

*Signature N**

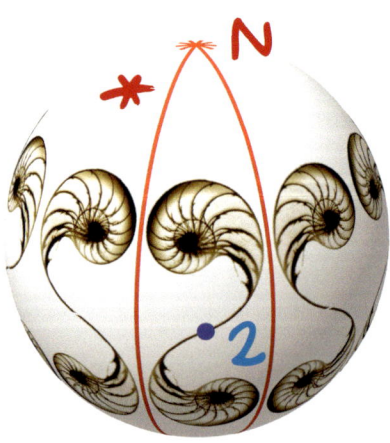

*Signature 2*N*

Signature Nx

The Existence Problem: Proving the Proviso

All 17 possibilities that we enumerated for planar patterns actually arose. In the spherical case, the corresponding statement is not quite true; the types **MN** and *MN only exist if $M = N$. The other cases cause no problem.

For example, *442 was generated by reflections in a triangle of angles $\frac{\pi}{4}$, $\frac{\pi}{4}$, $\frac{\pi}{2}$, and a planar pattern with this type of symmetry exists because such a triangle exists in the Euclidean plane.

Similarly, *532 is generated by reflections in a triangle of angles $\frac{\pi}{5}$, $\frac{\pi}{3}$, $\frac{\pi}{2}$, and a spherical pattern with this symmetry exists because there is a spherical triangle with these angles, drawn at right. Two of these triangles together form a fundamental region for **532**.

In the plane, the sum of the angles of a triangle is always π radians $= 180°$. On a sphere, the sum will always be larger.[‡]

For example, a triangle on the globe with one vertex on the North Pole and two vertices on the equator has two angles of $\frac{\pi}{2}$ radians at the equator. Since we may choose the angle at the pole, we can create a fundamental region for any *22N. Gluing two of these together, we have a fundamental region for **22N**.

Now for the proviso! The type *MN, when it exists, is generated by the reflections in the sides of a two-sided polygon with angles $\frac{\pi}{M}$ and $\frac{\pi}{N}$. This *does* exist when $M = N$; it's the lune bounded by two great semicircles at angle $\frac{\pi}{N}$ (at right), but does not when $M \neq N$. (For the same reason *M, which

equals *M1, fails to exist for $M > 1$.)

A hypothetical pattern of type **MN** with $M \neq N$ would contain just two types of gyration point. But then, by superposing it with its image under a reflection fixing a gyration point of each type, we should obtain one of type *MN, which is impossible. Therefore, **MN** also fails to exist if $M \neq N$, and **M** fails to exist if $M \neq 1$.

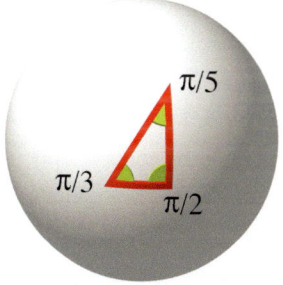

A triangle on the sphere with angles $\frac{\pi}{5}$, $\frac{\pi}{3}$, and $\frac{\pi}{2}$.

A two-sided spherical polygon with two angles of $\frac{\pi}{6}$ radians.

Group Theory and the Spherical Symmetry Types

A symmetry of a pattern is a transformation that does not change the pattern as a whole. The collection of all of the symmetries of a pattern form what mathematicians call a *group*. The central problem of this book is to classify the groups that can be the symmetries of planar and spherical patterns. We expect some readers will be surprised that we do this without using any group theory! This is because our

Magic Theorems rely on simple and powerful tools from topology, as we explain in the next few chapters. In the full edition of *The Symmetries of Things*, we discuss the symmetry groups of patterns in more detail, and there are many texts available that teach group theory better than we are able to in the space available here.

[‡] In fact, the area of the triangle is just this excess!

In brief, though, consider this gyroscopic pattern. It has exactly twelve symmetries: we may rotate by a twelfth clockwise, rotate by two-twelfths, etc., on up to rotating by eleven-twelfths clockwise. The twelfth symmetry — the "identity" — is the one that does nothing at all, or just the same, rotates by any number of full turns. These symmetries may be combined: we may first rotate by, say five-twelfths clockwise and then by nine; the end result would be the same as rotating by fourteen-twelfths, or, more simply just two, which (of course) is also one of our twelve symmetries.

Every one of the patterns in this book has a symmetry group, a collection of rigid transformations that do not change the pattern as a whole, with rules for how they may be combined.

Any pair of symmetries A and B of a pattern have a product AB, obtained by performing the motions of the pattern corresponding to A and B one after the other. Since the transformation A left the pattern unchanged and the transformation B left the pattern unchanged, doing one after the other also leaves the pattern unchanged, and their product AB is also a symmetry in the group.

If you think about it you may see that in the examples we have seen so far, this composition is associative (i.e., that $(AB)C = A(BC)$ for all choices of A, B, and C) and that in the relation $AB = C$, any two of A, B, and C uniquely determine the third.

The identity is special: when we combine the identity with any other symmetry, we don't change the result. And every symmetry has an "inverse" that is its undoing; combining a symmetry with its inverse produces the identity.

All together, these properties imply very many others, and are the definition of a group. The geometrical groups are the different ways that groups can act upon the plane, or sphere, or some other space, and the signatures are names for these groups.

Different geometrical groups have the same abstract structure if their elements multiply in the same way. For example, the geometrical groups with signatures 2• and ∗• both consist of two symmetries, the identity and another transformation, which is undone by repeating itself a second time. As abstract groups, these have the same multiplication table and are equivalent.

Table 4.1 lists the types of spherical repeating patterns and their equivalences as groups. The table is complete for patterns of order up to 24, after which we restrict to multiples of 3.

The last column gives the number of distinct geometrical groups, followed in parentheses by the number of different abstract structures (as separated by the lines). Groups with different abstract structures are separated by the curved and straight lines.

Codes for these structures are given below the table and are explained in Table 4.1: The polyhedral groups are isomorphic to the alternating and symmetric groups A_4, S_4, or A_5 according as n is 3, 4, or 5.

It may be surprising that the symmetry groups 432 and ∗332 have the same abstract structure. Below, we can see that 432 permutes the four axes of a cube, and thus the colors. Indeed, there's a perfect one-to-one correspondence between the permutations of the colors and the symmetries in 432, and they multiply together the same way too — 432 is isomorphic to the group of permutations S_4.

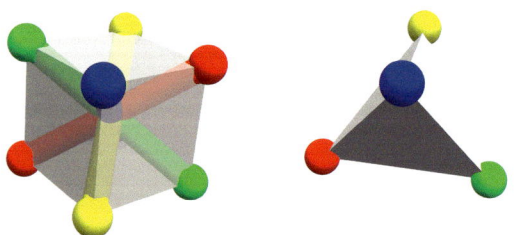

In the same way, each symmetry of ∗332 permutes the four vertices of a tetrahedron, and every permutation of the vertices determines a symmetry — ∗332 is also isomorphic to S_4, and thus to 432.

Order	NN	Nx	N*	2*N	22N	*NN	*22N			Number of Groups
1	11									1
2	22	x	*		22	*				3(1)
3	33									1
4	44	2x	2*	2*	222	*22	*22			5(2)
5	55									1
6	66	3x	3*		223	*33				5(2)
7	77									1
8	88	4x	4*	2*2	224	*44	*222			7(4)
9	99									1
10	10 10	5x	5*		225	*55				5(2)
11	11 11									1
12	12 12	6x	6*	2*3	226	*66	*223	332		8(4)
13	13 13									1
14	14 14	7x	7*		227	*77				5(2)
15	15 15									1
16	16 16	8x	8*	2*4	228	*88	*224			7(4)
17	17 17									1
18	18 18	9x	9*		229	*99				5(2)
19	19 19									1
20	20 20	10x	10*	2*5	22 10	*10 10	*225			7(3)
21	21 21									1
22	22 22	11x	11*		22 11	*11 11				5(2)
23	23 23									1
24	24 24	12x	12*	2*6	22 12	*12 12	*226	*332 432	3*2	10(6)
27	27 27									1
30	30 30	15x	15*		22 15	*15 15				5(2)
33	33 33									1
36	36 36	18x	18*	2*9	22 18	*18 18	*229			7(3)
39	39 39									1
42	42 42	21x	21*		22 21	*21 21				5(2)
45	45 45									1
48	48 48	24x	24*	2*12	22 24	*24 24	*22 12		*432	8(5)
51	51 51									1
54	54 54	27x	27*		22 27	*27 27				5(2)
57	57 57									1
60	60 60	30x	30*	2*15	22 30	*30 30	*22 15	532		8(4)
63	63 63									1
66	66 66	33x	33*		22 33	*33 33				5(2)
⋮	⋮	⋮	⋮	⋮	⋮	⋮	⋮			⋮
120	120 120	60x	60*	2*30	22 60	*60 60	*22 30		*532	8(5)
	C		$2 \times C$		D		$2 \times D$	P	$2 \times P$	

with

		for $n = 3, 4, 5$
C	cyclic	$\langle a \mid a^n = 1 \rangle$
D	dihedral	$\langle a, b \mid a^n = b^2 = (ab)^2 = 1 \rangle$
P	polyhedral	$\langle a, b, c \mid a^n = b^3 = c^2 = abc = 1 \rangle$
$2 \times G$	direct product of G with a group of order 2.	

TABLE 4.1. *Types of spherical repeating patterns and their equivalences as groups.*

Where Are We?

In this chapter we have shown that the spherical form of the Magic Theorem implies that the spherical symmetry types fall into seven infinite families plus seven individual types. The groups of symmetries of these patterns are listed by increasing number of symmetries in Table 4.1. After taking a look at many examples of spherical patterns in the rest of this chapter and frieze patterns in the next, we will reach the proof of the Magic Theorem in Chapter 6.

Examples and Exercises

The Regular Polyhedra

A polygon is called regular if all of its edges are of equal length and all of its angles are congruent. A *regular polyhedron* is a polyhedron with identical regular polygons for its faces, and the same number of faces at each of its vertices. On page 113 we will see why there are just five of these regular polyhedra (once we further presume they are topological spheres and not tangled up somehow in space). Below we show them, together with their names and signatures.

Each of the regular polyhedra has its own mirror planes of symmetry. At right we show these for a cube. In the figures below, mirror planes intersect the polyhedra in the red lines.

The regular polyhedra are related in dual pairs, sharing the same kaleidoscopic symmetry: At left below, the cube and octahedron both have symmetry *432. At right below, the dodecahedron and icosahedron both have symmetry *532. The tetrahedron is dual to itself, with signature *332.

You can verify the symmetry types of the regular polyhedra from the drawings, but it is much more fun and instructive to make your own models!

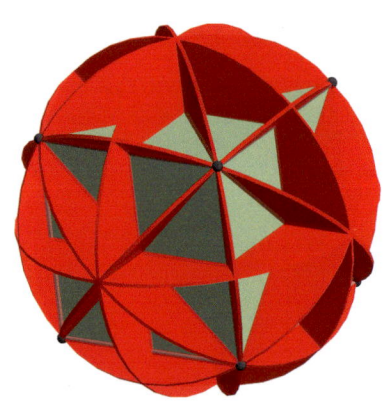

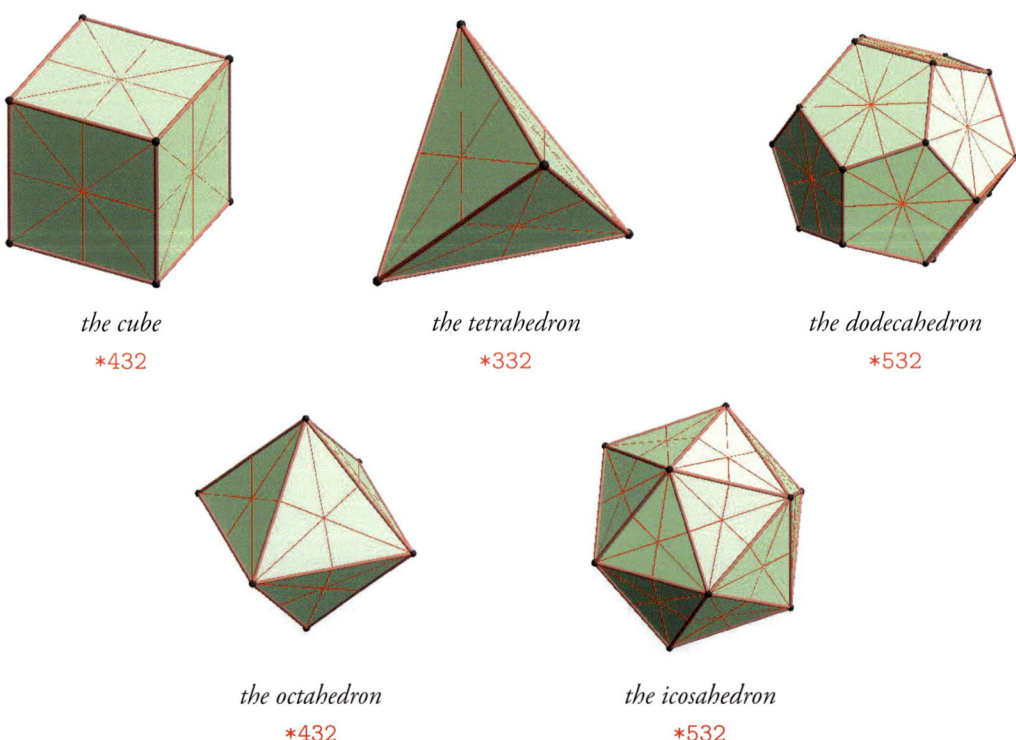

the cube

*432

the tetrahedron

*332

the dodecahedron

*532

the octahedron

*432

the icosahedron

*532

Eyeglasses. Each of these pairs of eyeglasses has two symmetries — the one that does nothing, and one that swaps the two lenses. The usual pair of eyeglasses at upper left has signature ∗ and a mirror symmetry, swapping from right to left. The unusual pair at upper right in the photo has an inversive symmetry, swapping points in opposite directions from the bridge, and its signature is ×. The two pairs at bottom both have signature 22, but their 2-fold rotation axes are oriented differently.

Symmetries of Playing Balls

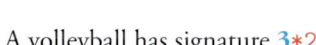

A volleyball has signature 3∗2

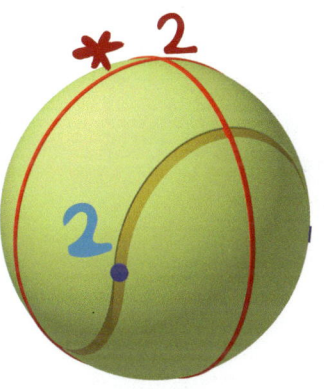

A tennis ball has mirror symmetry and rotations, with signature 2∗2.

A baseball has stitching, and signature 2x.

The most common kind of soccerball has signature ∗532. Turn to pages 76 and 77 for other types!

What signature does a basketball have? It might not be what you expect — Find one and take a look!

Polyhedral Models

Let's analyze the symmetry types of a few paper polyhedra. Among our teaching materials (page 168), we've put kits for you to download, print out, and assemble into your own polyhedral models.

An unmarked cube has signature *432. This cube has signature 332.

This faceted icosadodecahedron has signature *532.

The snub dodecahedron has signature 532.

The cube and octahedron, and their compound shown here, all have signature *432.

This marked octahedron has signature 3*2. Do you see the icosahedron hidden within it?

This compound of two dodecahedra has 3*2 symmetry.

Though a tetrahedron has symmetry *332, this decorated compound of two of them (a *stella octangula*) has the same symmetry as a cube, *432.

This regular tetrahedron has signature *332 and the regular icosahedron inside of it has signature *532. This model of the two together has signature 332.

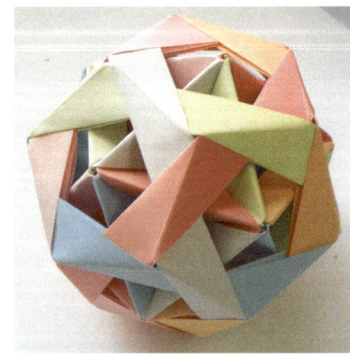

If we ignore the colors, this origami model has signature 532; the signature of each (one-colored) band is 225. The colored model has trivial signature •: no motion fixes all of the colors.

More Polyhedra to Analyze

Here are a few more polyhedra to analyze. You can check your signatures on page 82.

Look for 4-fold, 3-fold, and 2-fold axes. By the Magic Theorem for spherical patterns, what must the signature of this polyhedron be?

Identify the signature of this colored stellation of the snub cube.

One of these five tetrahedra has been specially painted — What is the symmetry type of this model? What would it be if all of the tetrahedra were identical? (Build your own for a closer look!)

By finding mirror planes you can determine the signature of this compound of three colored cubes. What if we ignore the colors?

Bathsheba Grossman's Sculptures

Grossman's sculptures reveal our lack of full intuition about three dimensional symmetry; the symmetry type can really only be appreciated by holding the model and examining it from several points of view.

The "Quintrino" has a five-fold axis of rotational symmetry and so is easily recognized as having signature **532**…

…which we can verify from other vantage points. Here we look down a two-fold axis of rotational symmetry.

We can guess that "Ora" has the rotational symmetry **332** of the tetrahedron.

Here is a view of Ora down a two-fold axis.

These are more difficult to recognize.

The "Soliton" is somewhat more mysterious. Here is a view down one 2-fold axis…

…and here is another. There is one more two-fold axis — which we don't show here — and the signature is 222.

The "Clef" also has signature 222, though it is hard to imagine from this one image how the views down the other axes might appear.

Can you guess what type this might have?

Temari balls are a fascinating way to realize spherical patterns.

This ball made by Ginny Thompson has gyration points of order 2 …

and a kaleidoscope *3; the signature is 2*3.

This ball has signature 2 2 12.

Another example of an object with cubic symmetry, signature *432.

This ball has signature 532.

At first glance, this has signature *22N for some large N. But if we pay very close attention to the weaving, the mirror symmetries are broken and the signature is 22N.

QUIZ: *What are the symmetry types of these beautiful temari balls, created by Carolyn Yackel? Their signatures are on page 82.*

Soccer Balls

QUIZ: *What are the symmetry types of Jon-Paul Wheatley's playing balls?*
Check your answers on page 83.

The Badly Drawn Ball.

After the Adidas Al-Rilha.

Another design, with 232 panels.

Using the "hat" monotile.

David Swart gathered these soccer balls in [16]. Every spherical symmetry type appears, some more than once.

Quiz: *How many signatures can you identify?*
(Ignoring stitching and logos, the signatures appear on page 83.)

Spherical Kaleidoscopes

Physical kaleidoscopes that generate spherical patterns are uncommon but worth the effort to build yourself. Below we include plans for their construction! The balls at the beginning of this chapter were raytraced in virtual kaleidoscopes; here are photos of real ones.

Not a kaleidoscope!

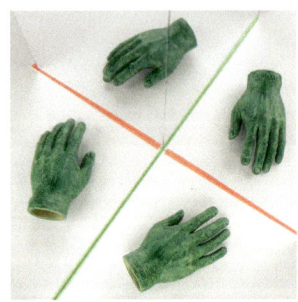

As a spherical kaleidoscope, this has signature *22.

Two mirrors will form a simple kaleidoscope with signature *NN if (and only if) they meet at an angle of $180°/N$, an even fraction of a circle. In a true mathematical kaleidoscope, like the one at far right, the images will always appear whole. When you look into mirrors meeting at any other angle, images will appear fractured behind the line where the mirrors meet. The mirrors in the photograph at close right meet at an angle of a fifth of a circle and do not form a true mathematical kaleidoscope.

Can you explain the "Mirror Paradox" that arises when $N = 2$? When peering into a pair of mirrors meeting at a right angle, like the ones in the photograph at the top right, your image is not reversed. The real toy is left-handed and its image directly across in the back of the kaleidoscope is too.

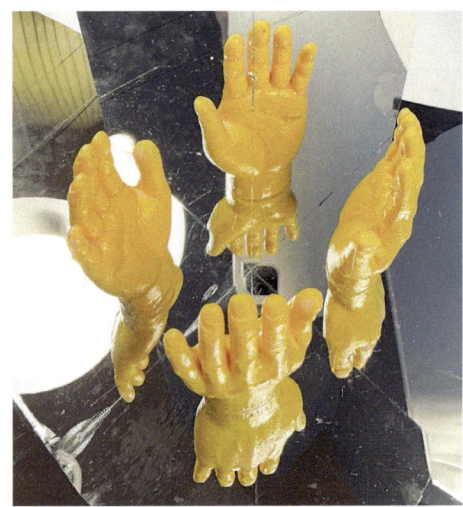

A di-scope, with signature *222.

The di-scope, right. For a slightly more elaborate kaleidoscope with signature *2NN, rest a pair of mirrors meeting at an angle of $180°/N$ on a horizontal mirror. At right is *di-scope* with all the mirrors meeting at right angles and signature *222. Which of these toys are left hands and which are right hands?

A tetrascope with signature 332.

The Tetrascope, with signature *332: Cut mirrors as shown below and fold into a cone. Drop objects into the chamber to see images with this signature. With special blocks you can show polyhedra with this symmetry, like the stella octagula at left.

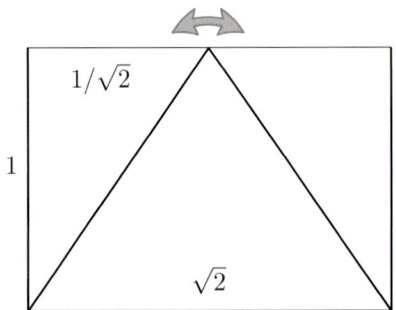

The Octascope creates patterns with signature *432. At right we show a block shaped like 1/48th of a cube, placed within an octascope. With the right pieces, we could show an octahedron, or any other polyhedron with this symmetry.

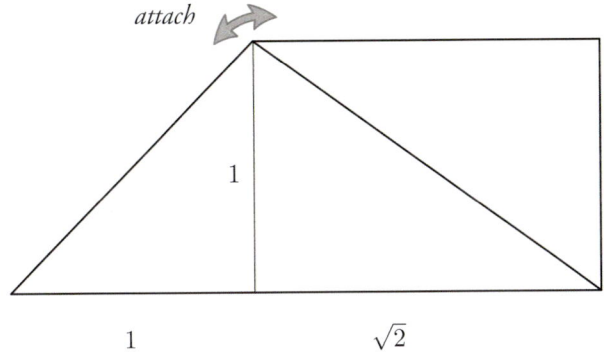

An octascope with signature 432.

The Icosascope with signature *532 is the most marvelous of all! Cut the mirrors as shown to the right and fold into a cone. At left, smaller cardboard model of the same shape is put into an icosascope to show that it is really 1/120th of a dodecahedron. If you cut a hole on one end, along the gray lines, you will see a pattern in the shape of a stellated dodecahedron!

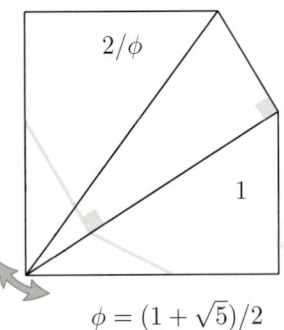

$$\phi = (1 + \sqrt{5})/2$$

An icosascope with signature 532.

Another icosascope, with the same signature.

Generating Regular and Archimedean Polyhedra in Kaleidoscopes

The regular polyhedra are related in dual pairs, sharing the same kaleidoscopic symmetry. On page 68, we see that the icosahedron and dodecahedron both have symmetry type *532; the cube and octahedron both have *432 symmetry. The tetrahedron is dual to itself, with *332 symmetry.

In fact, for any of these pairs, there's a continuum of polyhedra between them, all with the same symmetry, shown below for the dodecahedron and icosahedron pair, drawn in

Jeff Weeks' *Kaleidotile* software. Each member of this family is given by a point in the kaleidoscope, known as a *Wythoff triangle*.

These are some of the *Archimedean* polyhedra. (On page 70 we show an Archimedean polyhedron that has 532 symmetry, and so doesn't have a kaleidoscope. See Chapters 19 and 21 of the full edition of *The Symmetries of Things* for a complete accounting.)

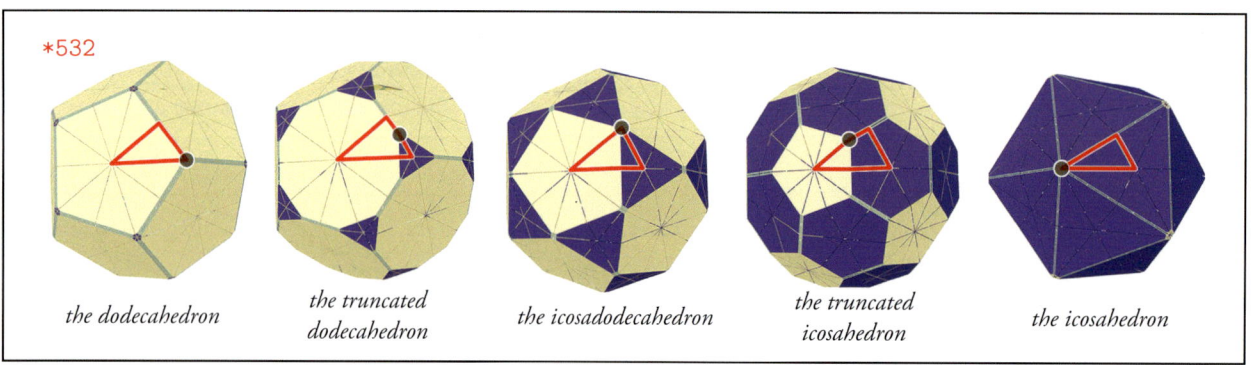

*532

| the dodecahedron | the truncated dodecahedron | the icosadodecahedron | the truncated icosahedron | the icosahedron |

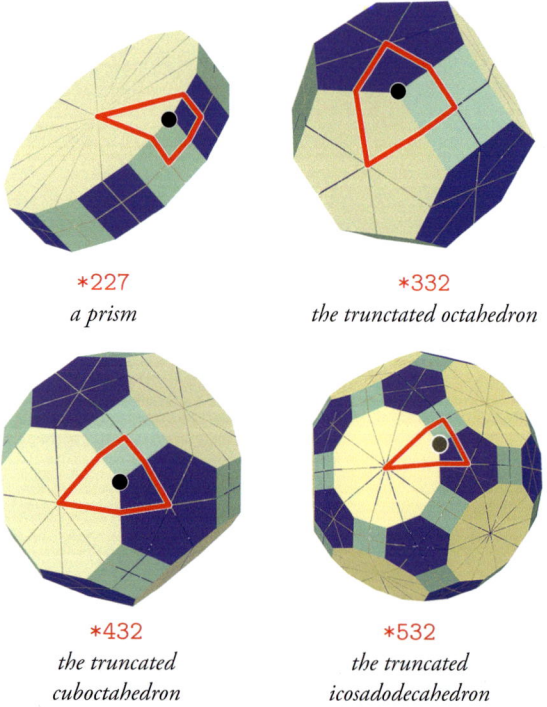

*227
a prism

*332
the truncated octahedron

*432
the truncated cuboctahedron

*532
the truncated icosadodecahedron

The same construction works in any symmetry of type *PQR — this is the unity of the orbifold perspective. The figures at left show the Archimedean polyhedra corresponding to the center of the Wythoff triangle, in various kaleidoscopic symmetries. Below at left we show a planar tiling with signature *632. The construction seamlessly carries over into hyperbolic space, as shown below right for *732 (Chapter 10).

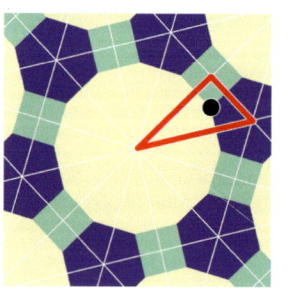

*632

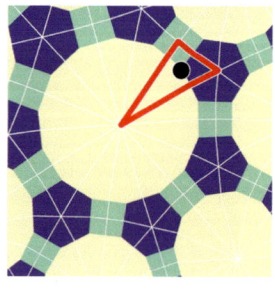

*732

Unusual Dice

Top row: 5-, 6-, 7-, 9-, and 11-sided dice. Second row: 12-, 14-, 14-, 18-, and 22-sided dice. Third row: 24-, 24-, 30-, and 120-sided dice.

Here are some interesting dice, produced by Dice Lab, Kosmo Games, and Impact Miniatures, with unusual shapes and numbers of sides. Some of these dice are not really fair, and some don't look fair but are!

A die will be unbiased if there is a symmetry taking any particular face to any other, so that each face is as likely to come up as any other [6, 9]. We may reasonably presume the converse, that a die must be biased, at least subtly, if it has different kinds of faces up to symmetry.

The symmetry types of dice can be difficult to work out from a photograph, but we can find some clues. For an unbiased die, the number of its faces must divide the order of its symmetry group. From Table 4.1 we can see that if any of our dice with 5, 7, 9, 11, 14, 16, or 22 sides were fair,

it would have a symmetry type in one of the seven infinite families **NN**, **22N**, **∗NN**, **∗22N**, **2∗N**, **N∗**, and **NX**. Things that have these signatures all have one special axis, about which we may rotate by $1/N$ of a revolution. The red 14-sided d7 at center does have such an axis, and has signature **227** — it's fair. None of the rest of those dice has a suitable axis, and none are fair. In fact, many of these are Archimedean polyhedra, and clearly have more than one kind of face.

On the other hand, the 6- and 12-sided dice in this photograph do not look like they are fair, but they are! Their signatures are **322** and **332**. The dice on the bottom row are all fair, with one kind of face in the symmetry — the dice are *isohedra*, two with signature ∗432, two with signature ∗532.

Answers:

The signature of a basketball (p. 69) is 2∗2. Top row, page 71: From left, 432; 332 accounting for colors, 432 if not; 332 as colored, 532 if not colored, • if the five tetrahedra are colored differently; ∗222 as colored, 3∗2 if uncolored. See page 168 for designs to print out and assemble for yourself!

Second row, page 75, overlooking finer details: 432; ∗332; 532; and ∗532. Next are the signatures of the objects with spherical symmetry on page 9: These subtle fair dice from Dice Lab have signature 322; Dick Esterle's Knobbly Wobbly has signature • accounting for the colors, 532 if not; Bathsheba Grossman's Clef has signature 222; the Craighill Jack Puzzle has signature 3∗2; John Kostick's brass star has signature 532. Shiying Dong's crocheted Seifert surface has signature 332 if the colors are not considered. Accounting for them, it has signature • — to preserve the colors, the only possibility is do nothing.

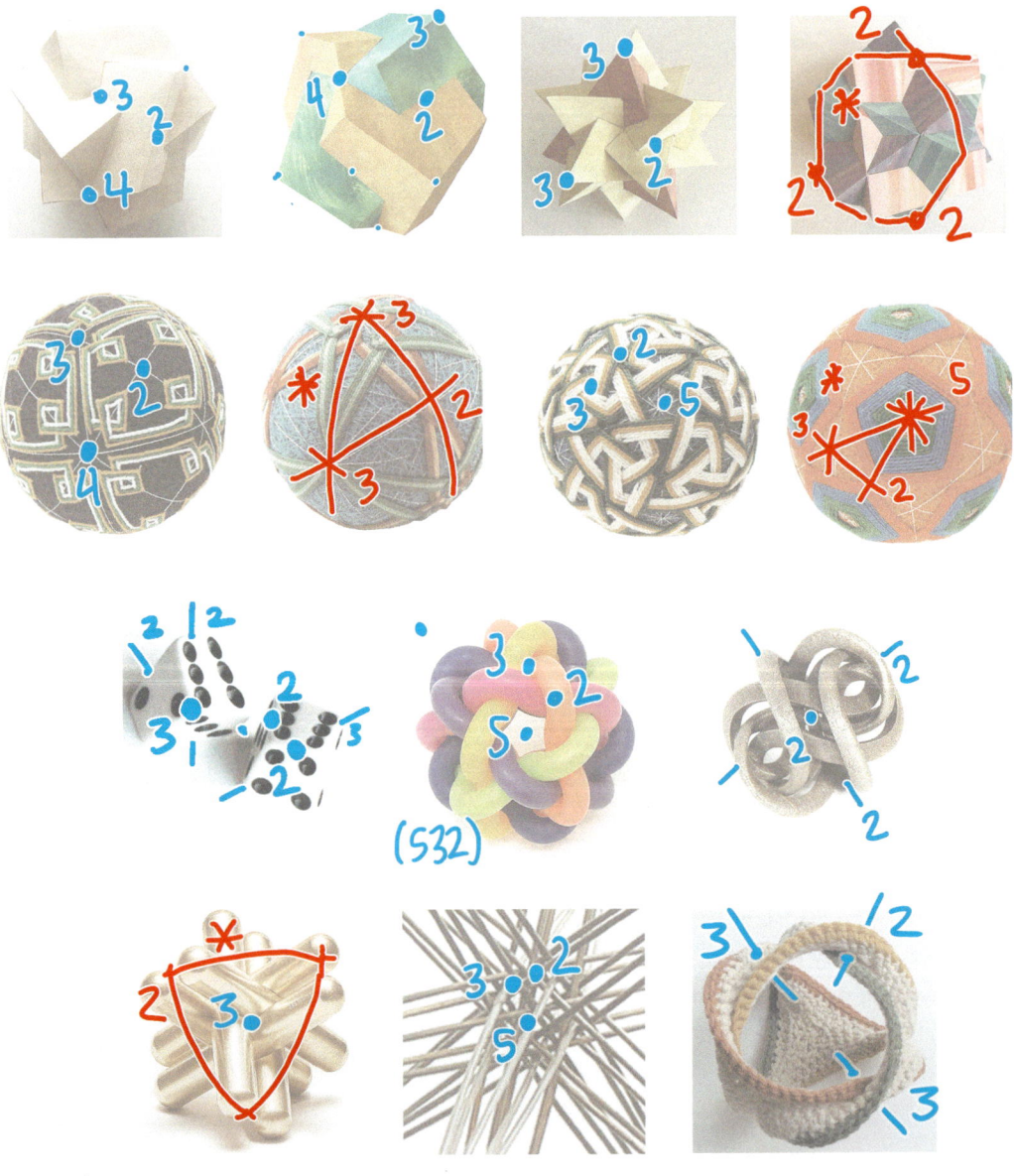

The signatures of the balls on pages 76 and 77: Top row: The Badly Drawn Ball has trivial symmetry, •. Al-Rilha has signature 3*2. The 232 Panel Ball first appears to have signature 532, but on closer inspection has trivial symmetry. The hat ball has signature 532, and is based on a planar periodic tiling with hexagonal holes and signature 632.

Second row: Eigil Nielsen's iconic Telstar ball (1968) has signature *532; the other balls have 222 and 2*2. Third: 3×; *432; *532; *332. Fourth: 332; 432; 3*2. Fifth: 33; 2*; *55; 532. Sixth: *222; *432; the ball decorated by flags has trivial symmetry, signature •.

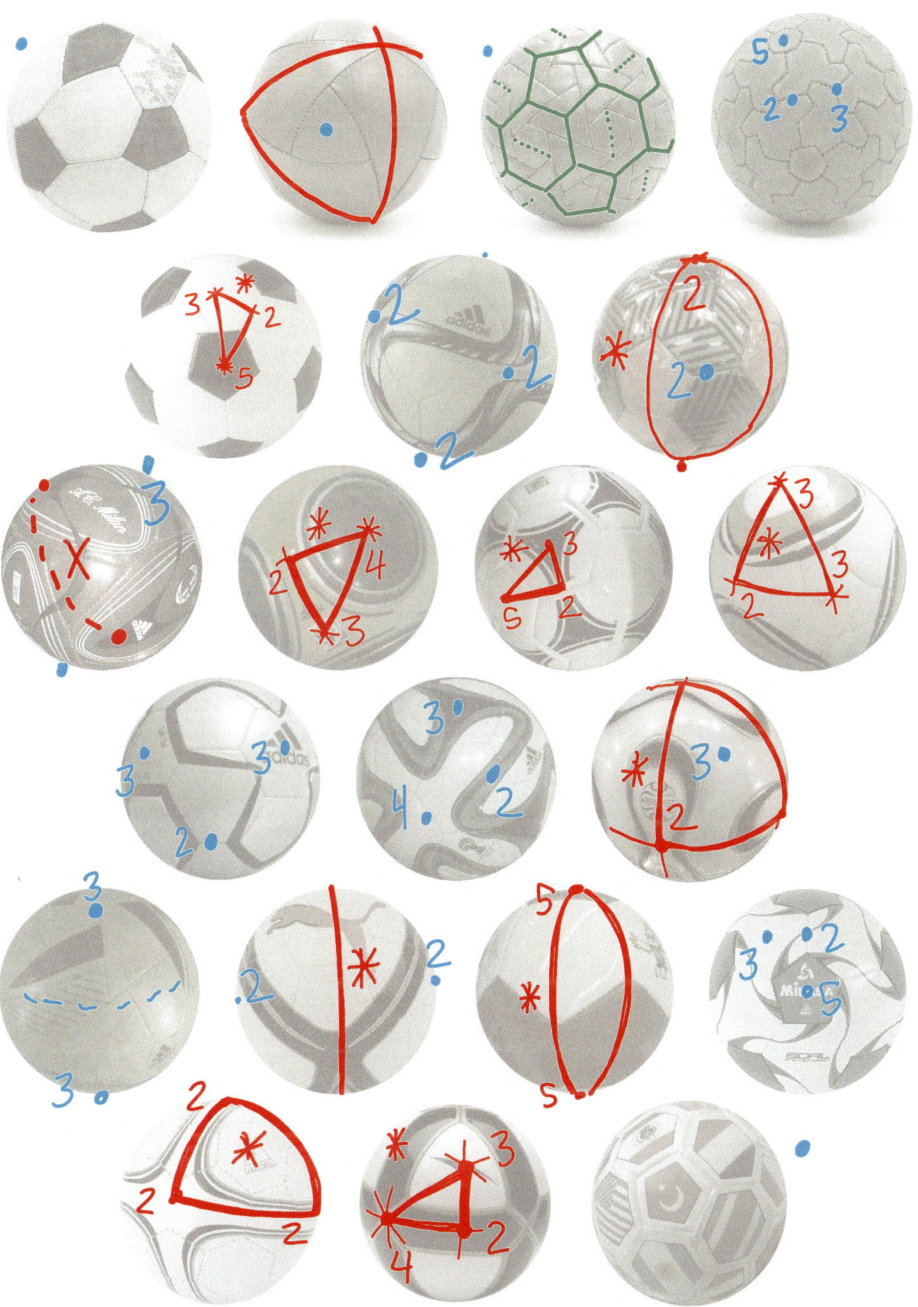

Chapter 5
The Seven Types of Frieze Patterns

There are other interesting patterns we've not yet considered. They are formed by the symmetries of plane patterns that repeat infinitely in one direction only: We call them *frieze patterns*. The facing page shows the seven different types of frieze pattern.

As we shall see in a moment, there is a Magic Theorem that we can use to list these. However, we don't really need it because any frieze pattern can be wrapped around a finite object such as a vase, which means that you can find the signature for a frieze pattern just as we did in the Euclidean and spherical cases:

Imagine the pattern wrapped around the equator of a very big sphere, such as in the figure at the top of this page.

According to the number of repetitions of the fundamental region, this wrapped up frieze pattern will have one of the seven spherical symmetry types that involve a parameter N — namely **NN**, **N×**, **N∗**, **∗NN**, **22N**, **∗22N**, or **2∗N** — and so it's natural to say that the corresponding infinite frieze pattern has symmetry type $\infty\infty$, $\infty\times$, $\infty*$, $*\infty\infty$, 22∞, $*22\infty$, or $2*\infty$. These could also be deduced from the following.

Theorem 5.1 (The Magic Theorem for Frieze Patterns) *The signatures of frieze patterns are precisely those that contain an ∞ symbol and cost exactly* $^\$2$.

The symbol ∞ costs $^\$1$, which makes perfect sense since $\frac{\infty-1}{\infty} = 1$. The symbol ∞ costs $^\$\frac{1}{2}$, since $\frac{\infty-1}{2\infty} = \frac{1}{2}$.

(opposite page) Seven frieze patterns to analyze, with signatures on page 94.

hop: ∞∞

step: ∞ ×

jump: ∞∗

sidle: ∗∞∞

dizzyhop: **22**∞

dizzyjump: ∗**22**∞

dizzysidle: **2**∗∞

FIGURE 5.1. *Tripping around the world in seven different ways!*

Figure 5.1 shows frieze patterns formed by footprints in the sand of an infinite desert plane. To analyze them, we transfer each one to a finite spherical planet, such as the one at right. There our previous methods show the resulting signatures of these wrapped up footprints to be **NN**, **N×**, **N∗**, ∗**NN**, **22N**, ∗**22N**, and **2∗N** for very large N. (The pattern of footsteps on the sphere at right has signature **11 ×**.) The originals were therefore ∞∞, ∞×, ∞∗, ∗∞∞, **22**∞, ∗**22**∞, and **2∗**∞, respectively.

The patterns for the types **NN**, **N×**, **N∗**, and ∗**NN** are what we get when we hop, step, jump, or sidle around the world. For the types **22N**, ∗**22N**, and **2∗N**, we spin between each hop, jump, or sidle, so we call these the "dizzy" types, or "ditypes."

With a little practice, the types can be found directly from the original patterns. For instance, the "dizzy jump" or "dijump" pattern has the mirror lines:

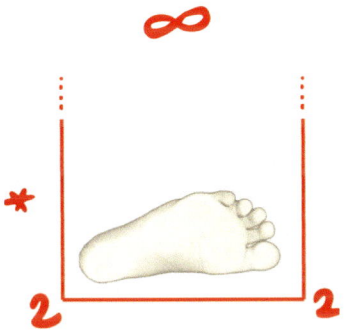

It is clear that the kaleidoscopes in this pattern have corners with angles of $\pi/2$ and that the kaleidoscopic points are 2-fold. We declare that the parallel sides of the kaleidoscope meet "at infinity" with an angle of π/∞, and so the signature of this pattern is ∗**22**∞. In a similar way, the infinity symbols in the signatures of frieze patterns refer to translations (regarded as rotations about the infinitely distant poles).

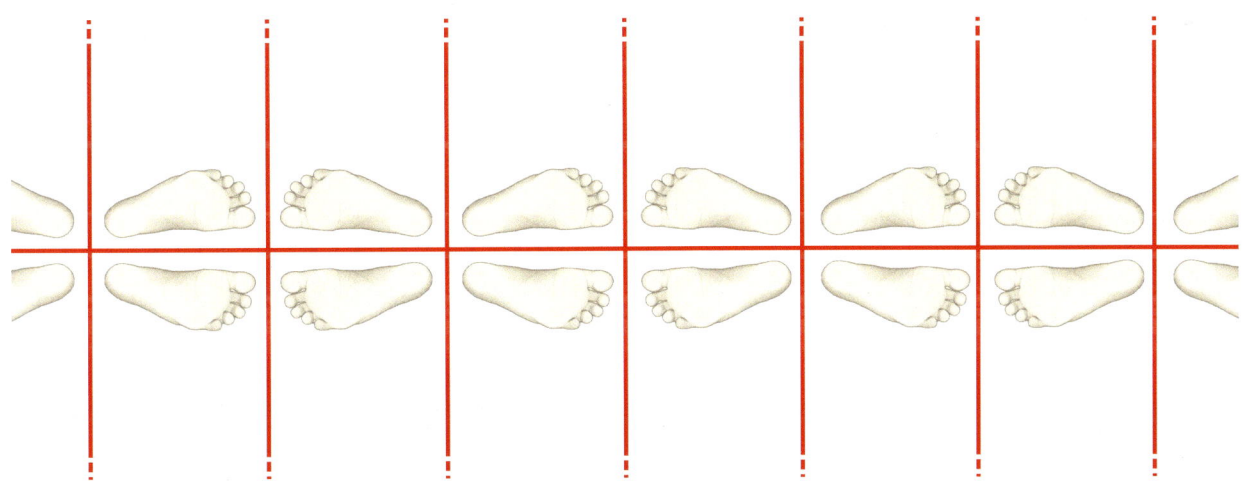

In just the same way, we can explain a hall of mirrors trick, below at right. Two mirrors meeting at an angle of π/N form a kaleidoscope with spherical symmetry type *NN. As the angle between the mirrors decreases, the number of symmetries (R's) in the pattern increases as seen in the top two figures.

Below these two figures, we see that as N increases without bound and the angle between the two mirrors decreases to zero, the pattern of R's becomes a frieze. The mirrors are now parallel and we can imagine that their ends meet infinitely far away at an angle of π/∞, at kaleidoscopic points of infinite order. The mirrors bound an infinite strip that repeats infinitely in both directions. The signature of this pattern is *∞∞. With two mirrors you can try this for yourself!

We can also view this pattern as the limit of a series of patterns with signature *N•. The symbol • in *N• is a stand-in,

indicating a kaleidoscopic point of order N sitting "at infinity". In fact, the spherical symmetry type of a rosette pattern with signature *N• is *NN.

Where Are We?

In Chapters 2–5, we've determined all possible types of symmetry for plane repeating patterns, spherical patterns, and frieze patterns using various forms of our Magic Theorem. So, the Magic Theorem is quite powerful.

What we haven't done is explain why it is true! Because this theorem is so powerful you might think it would be hard, but the next chapter shows that, in fact, it's quite easy.

Examples and Exercises

We list our answers on pages 94-95.

1. What are the signatures of the friezes shown on page 7?
2. What of those on page 84? (Their orbifolds appear on page 145).
3. What types are these alphabet friezes? Make up a few more! (What frieze symmetry types are possible with each letter?)

ZZZZZZZZZZZ

DDDDDDDDD

JJJJJJJJJJJJJJJJ

XXXXXXXXXXX

WWWWWWWW

W	D	Z	X	J
W	D	Z	X	J
W	D	Z	X	J
W	D	Z	X	J
W	D	Z	X	J
W	D	Z	X	J
W	D	Z	X	J

pppppppppppp

p d p d p d p d p d p

pqpqpqpqpqp

p b p b p b p b p b p

bbbbbbbbbbbbbbbbbbbbb
ppppppppppppppppppppppppppppp

bdbdbdbdbdbdbdbdbdb
pqpqpqpqpqpqpqpqpqp

bdbdbdbdbdbdbdbdbdb
qpqpqpqpqpqpqpqpqpq

4. Find the types of these coffee friezes.

5. Analyze the appearances of this Sonny Bono look-alike. Be careful: some of
these friezes have the same type and not every type is represented.

6. What are the symmetry types of these beautiful frieze patterns? Of course, more than one type may appear in the same photograph!

7. Here are some vertical friezes we spotted. What are their types?

8. A few more to practice on.

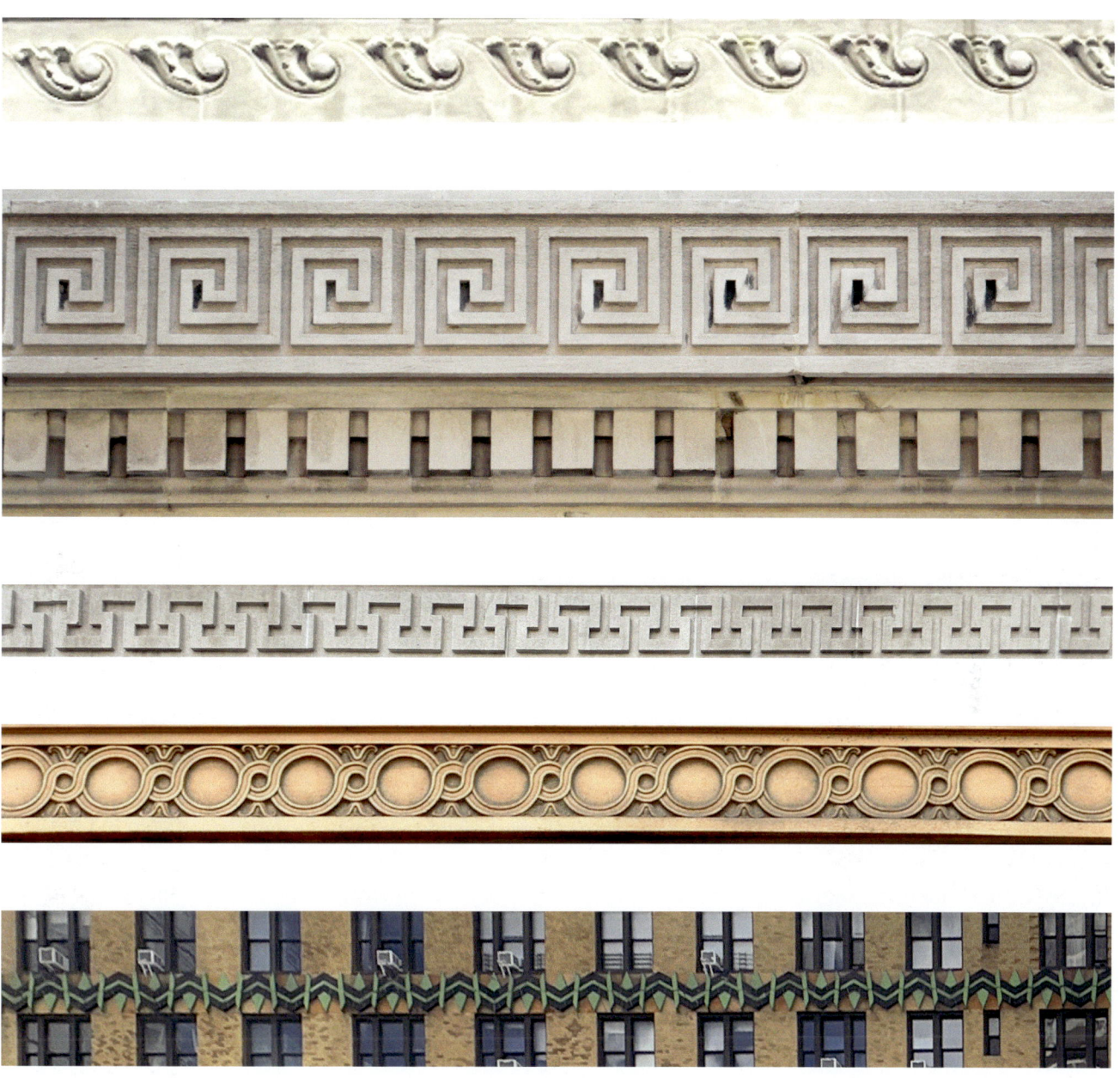

9. There are marvelous friezes in the world around you: What are their types?

Answers

1. From top to bottom, the friezes on page 7 are of types **22∞**, ∞x, **2∗∞**, **∗22∞**, **22∞**, and ∞∞∞. (It can help to imagine these on an enormous sphere with gyration or kaleidoscopic points of infinite order at its poles.)

  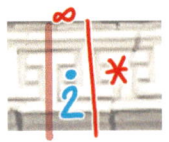 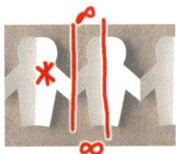  

2. The frieze patterns on page 84 are of types ∞×, ∞∗, ∞∞∞, **2∗∞**, **22∞**, **∗∞∞∞**, **∗22∞**. (On page 144, you can see how these patterns' signatures record the building blocks of their orbifolds.)

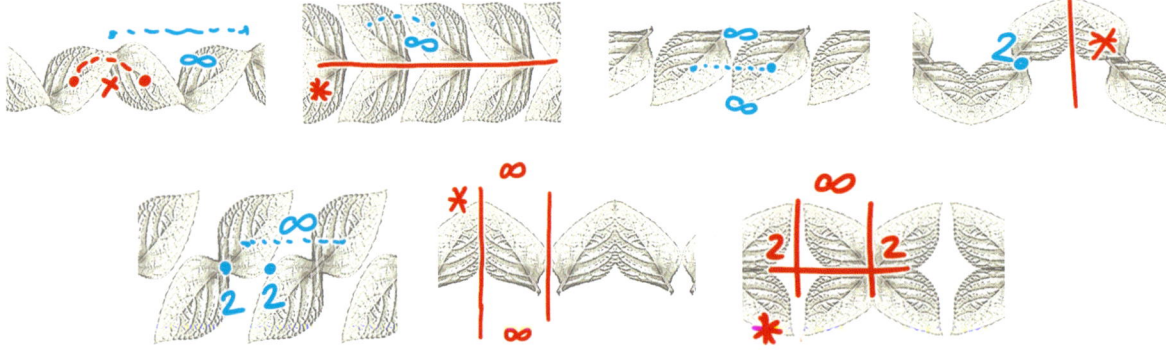

3. When repeated in a horizontal or vertical line, a letter with ∗2•point symmetry, such as H, I, X, or O (page 11), will form friezes with signature **∗22∞**. Letters with a single vertical mirror line, such as W,A,V,Y,T,U, or M will form horizontal friezes of type **∗∞∞∞** and vertical ones of type ∞∗. The reverse holds for letters with a single horizontal mirror, B, D, E, C, K. A letter with gyrational point symmetry **2•**, such as N, Z, or S, will produce a frieze of type **22∞** when it is repeated in any direction. The repeated p's have signature ∞∞∞; pd's have signature **22∞**; pq's **∗∞∞∞**; pb's **2×**. The remaining friezes have signatures ∞∗, **∗22∞**, and **2×**.

4. Listed in order from top to bottom: ∞∞∞, ∞×, ∞∗, **∗∞∞∞**, **22∞**, **2∗∞**, **∗22∞**.

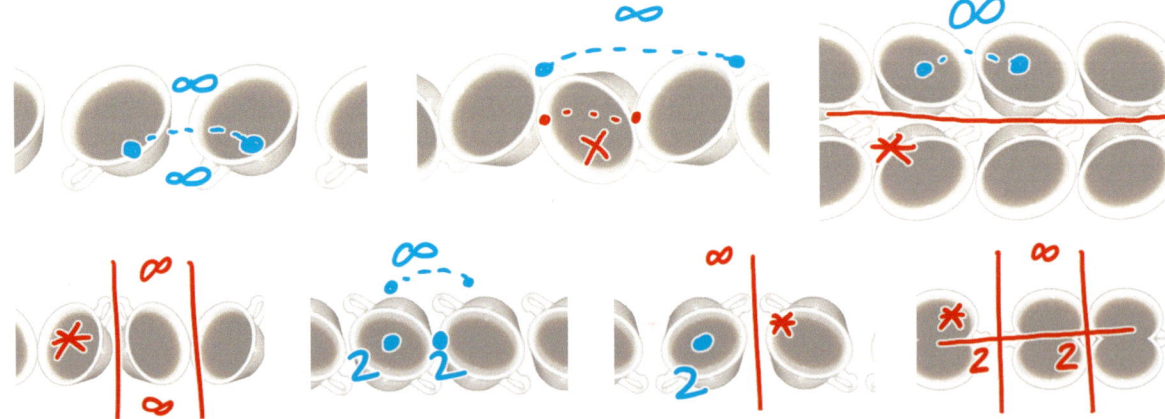

5. Listed in order from top to bottom: ∞∞, **2∗∞**, **22∞**, ∞∗, ∞×, ∗22∞, **2∗∞**.

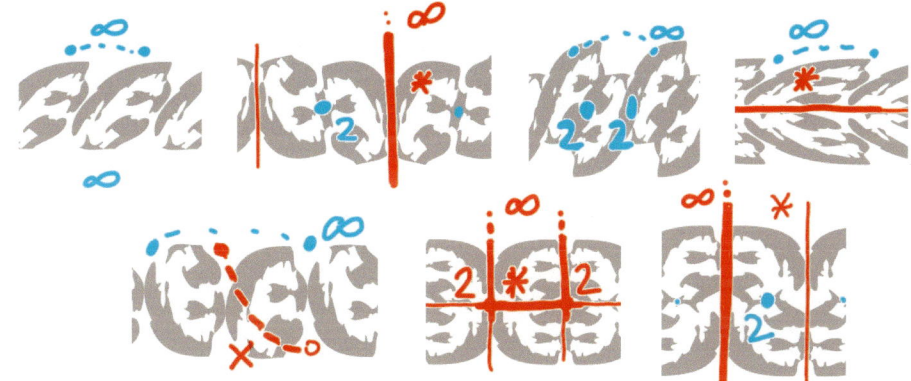

6. From top to bottom: First frieze **22∞**, ∗∞∞∞, and ∞x; second ∞x; third and fourth ∗∞∞∞. The fifth frieze shows types ∗22∞, ∗∞∞∞, and **22∞**. Sixth is ∞x and seventh is ∗∞∞∞.

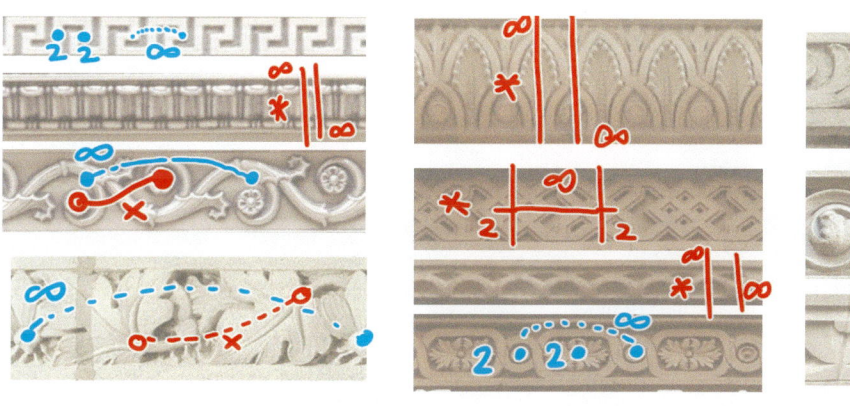

7. From left to right, **2∞**, not paying close attention to the screws; ∞∞∞ (if we may flip the stairs over, then **22∞**); ∞×; ∞∞∞; ∞×; ∞×.

8. From top to bottom: ∞∞∞; **22∞** and ∗∞∞∞; **22∞**; **2∗∞**; **2∗∞** (ignoring windows).

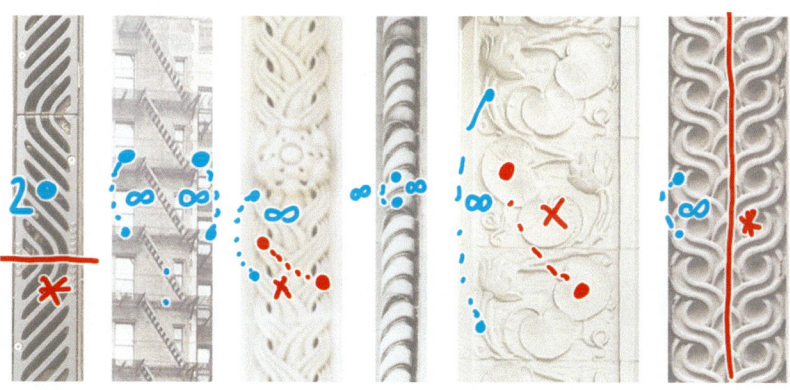

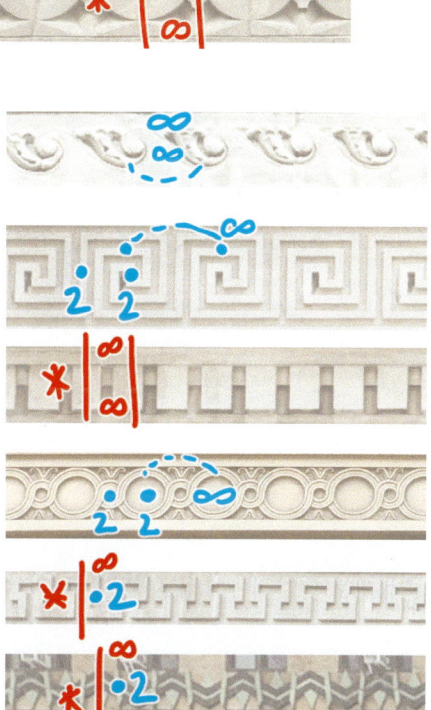

9. You tell us!

Chapter 6
Why the Magic Theorems Work

In this chapter we'll deduce the Magic Theorems from Euler's well-known theorem about maps. A mathematical map is like an ordinary map of countries and their borders. We'll show how the different features of a symmetric pattern affect the structure of some specially chosen maps and how Euler's theorem is used to determine the costs assigned to the features of a signature.

Folding Up Our Surface

We've told you that when several features in a pattern are of the same kind you should count them only once. We are really counting things not on the original surface but on a folded-up version of it, the folding taking all the points of the same kind to a single point. An *orbit* is a set of all of the points that are the same kind in a pattern, so this "orbit-folded" version of the surface is called the *orbifold*.

As we remarked in Chapter 4, the symmetries of finite objects can be thought of as symmetries of the surface of a celestial sphere. For example, the chair at the top of the page has two symmetries: the trivial one and the reflection in its plane of symmetry. This reflection equates pairs of points in the left and right of the celestial sphere, defining orbits. For example, the reflection equates the pair of blue points, and the pair of blue points is an orbit. The single red point lies on the mirror and is an orbit by itself. We can fold each orbit into a single point by pushing the right celestial hemisphere into the left one.

The orbifold is therefore a hemisphere. Most points of the orbifold, like the blue, green, and yellow points, correspond to full-sized orbits (of two points), but the boundary of the orbifold consists of half-sized orbits like the red one. The signature for this pattern is ∗ and its cost is $1.

At left, a planar pattern lifted stereographically up to a sphere.

Each pattern has an orbifold of its own, equating points that are of the same kind: the points in the orbifold correspond to orbits in the pattern. These orbifolds are topological surfaces, with a few marked points. The signatures we've been using describe topological properties of these surfaces. For example, the orbifold of a pattern of signature ∗ is a hemisphere, which topologically is a sphere with a very large hole ∗ punched into it.

In this chapter we will use topological tools to study orbifolds and prove our Magic Theorem, explaining why, for example, each hole costs $1, and why planar signatures total exactly $2.

Orbifolds

Each of the symbols in a signature describes a distinct feature of an orbifold. For example, around an N-fold gyration point marked **N** in a pattern, there will be N copies of each kind of point. On the orbifold we bring points of the same kind together until we have $1/N$th of a disk's-worth of material, forming a cone. We call such a point an N-fold *cone point*, and mark it as **N** on the orbifold. This crossword pattern has a 2-fold gyration point and its orbifold has a 2-fold cone point, which we mark **2**. (We take this up further on page 134.)

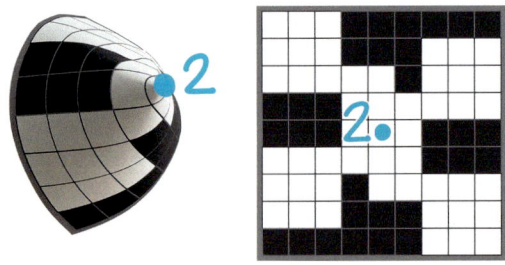

Near an N-fold kaleidoscopic point, an orbifold is a folded wedge, $1/N$th of a folded-over disk, called an N-fold *kaleidoscopic corner*.

This paper snowflake pattern has signature ∗7•. Folding this pattern along mirror lines brings points of the same kind together. Its orbifold has a 7-fold kaleidoscopic corner 7 on a piece of a boundary ∗. You can confirm that this is the orbifold by framing it between a pair of mirrors, or by cutting out your own paper snowflake from this design. (More of these are on page 132.)

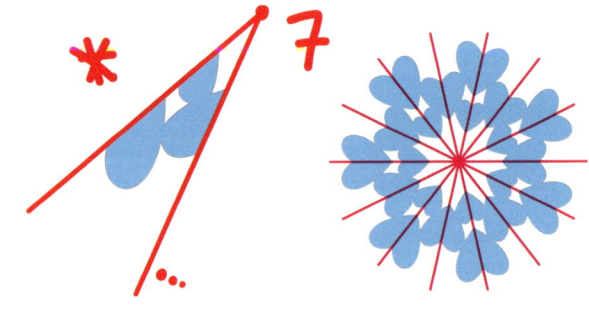

In a signature, a red digit N can only come after a ∗, because an N-fold corner can only lie on a boundary of the orbifold. For example the signature ∗632 describes an orbifold with a single boundary and three marked corners upon it, a triangle consisting of one copy of each of the different kinds of point in the pattern. (More kaleidoscopic orbifolds appear on page 133.)

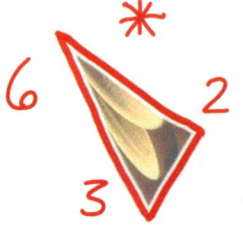

In Chapter 8 we will study topological surfaces, learning that they can all be described as a sum of "boundaries" ∗, "handles" ○, and "crosscaps" ×. Together with marked cone points **N** in the interior of an orbifold and corners **N** on its boundaries ∗, *these* are the features that the signature records. In Chapter 9 we'll take a closer look at some patterns and get a better feel for their orbifolds. For now we will use them to prove our Magic Theorems.

Euler's Map Theorem on Spherical Orbifolds

Leonhard Euler discovered a wonderful fact about maps drawn on a sphere — namely that $V - E + F = 2$, where V, E, and F are the numbers of vertices, edges, and faces of the map, respectively. We'll use *char* for $V - E + F$ since this number is traditionally called the *Euler characteristic*. The proof of Euler's Theorem is postponed to Chapter 7: For now we study what happens to *char* when we fold up maps on spheres into orbifolds.

At right, we show a sphere decorated with five vertices and eight edges. These divide the sphere into five separate regions, or faces of this map. So in accordance with Euler's Map Theorem,

$$char = 5 - 8 + 5 = 2.$$

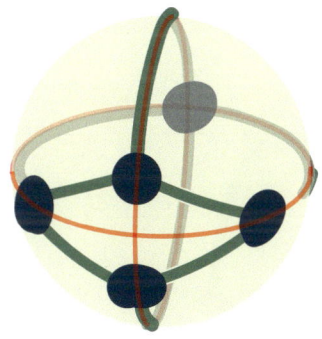

A map that has V = 5, E = 8, F = 5, and char = 2.

As shown by the red mirror lines this map has signature *22. Its orbifold is a quarter sphere, with one boundary *
and two corners 22 upon it.

On this orbifold, we see a folded form of this map. Some of the vertices, edges, and faces have been halved or quartered. Examining the figure at right, we have

$$V = \tfrac{1}{2} + \tfrac{1}{2} + \tfrac{1}{4} = \tfrac{5}{4},$$
$$E = 1 + \tfrac{1}{2} + \tfrac{1}{2} = 2,$$

and

$$F = 1 + \tfrac{1}{4} = \tfrac{5}{4}.$$

This quarter-spherical folded map has

$$char = V - E + F = \frac{5}{4} - 2 + \frac{5}{4} = \frac{1}{2}.$$

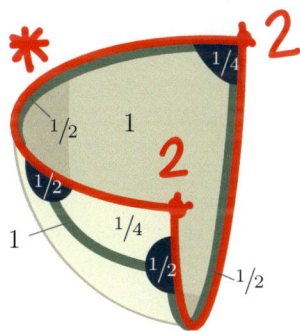

A folded map with char = 1/2.

The same argument works for any spherical map: When we bring together points that are the same in the pattern's symmetry, the resulting orbifold map has $char = 2/g$, where g is the number of symmetries of the pattern.

For example, the cubical map at left has forty-eight symmetries (taking any marked triangle to any other). When we fold this map along the mirror lines shown, the orbifold is a triangle ∗ with corners, with signature ∗432.

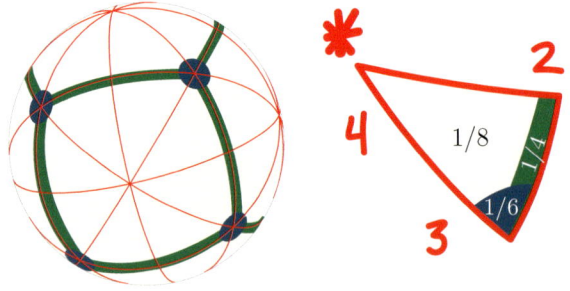

On its orbifold the map has only $\frac{1}{6}$ of a vertex, $\frac{1}{4}$ of an edge, and $\frac{1}{8}$ of a face, so

$$char = \frac{1}{6} - \frac{1}{4} + \frac{1}{8} = \frac{1}{24} = \frac{2}{48}.$$

This is obvious because all we've done is take 1/48th of $V - E + F = 8 - 12 + 6 = 2$ for the original cube map.

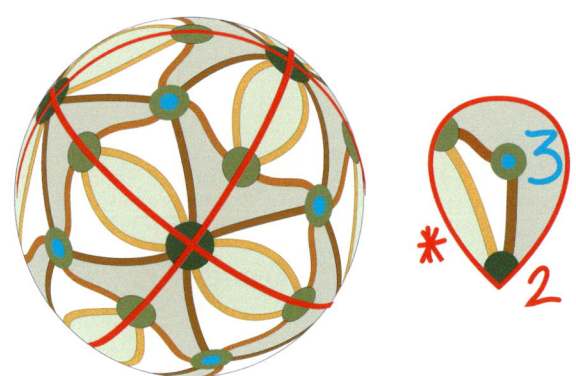

Similarly, this colorful map has signature 3∗2 — its orbifold has a 3-fold cone point 3 and a corner 2 on a boundary ∗. You can check that on this orbifold, we have

$$V = \frac{1}{4} + \frac{1}{3} + \frac{1}{2}, \quad E = 1 + 1 + 1, \quad \text{and} \quad F = 1 + \frac{1}{2} + \frac{1}{2}$$

giving $char = V - E + F = \frac{2}{24}$, which is 1/24th of the *char* of the original map on the sphere.

Why char = ch: *Proving the Magic Theorem for the Sphere*

We've now shown that for spherical types $char = 2/g$, so to prove the Magic Theorem in the spherical case we only need to explain why $char = ch$, the change after subtracting the cost of our signature from $^\$2$.

We work out how *char* changes as we add features to the orbifold — this is the cost of each feature. By Euler's Map Theorem, which we will prove in Chapter 7, any planar map has $char = 2$. To work out the orbifold Euler characteristic of a map like the one at upper right on the next page, we will choose a convenient planar map, like the one to its left, where the kaleidoscopic boundaries are filled in with faces, and the gyration and kaleidoscopic points are included among the map's vertices. Now we proceed to put in the features of the orbifold, to see the effect on *char*:

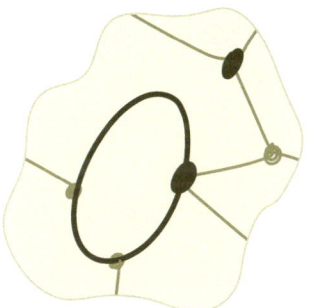 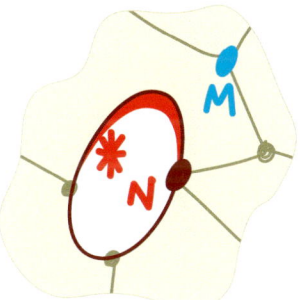

Adding features changes the cost.

Punching a boundary ∗ decreases char by 1. Choose a map for which the boundary is filled in by a single k-sided face, like in the drawing above. Then, removing it decreases F by 1 and V and E by $\frac{k}{2}$ (since vertices and edges around the hole get halved). Therefore, $V - E + F$ is reduced by $\frac{k}{2} - \frac{k}{2} + 1 = 1$; a boundary decreases *char* by 1, the cost of each ∗ in the Magic Theorems.

Replacing an ordinary point by an N-fold cone point **N** *decreases char by* $\frac{N-1}{N}$. Choose a map for which the point is a vertex. Before the change, it contributes 1 to V; afterwards it contributes only $\frac{1}{N}$. Therefore in the Magic Theorems the cost of an N-fold cone point is:

$$1 - \frac{1}{N} = \frac{N-1}{N}$$

Replacing an ordinary boundary point by an N-fold corner point N *decreases char by* $\frac{N-1}{2N}$. We choose a map for which the point lies on a boundary ∗; thus this point counts as half of a vertex. Replacing it with an N, it will be just $1/(2N)$ of a vertex, at a cost to *char* of

$$\frac{1}{2} - \frac{1}{2N} = \frac{N-1}{2N}$$

The orbifolds for 13 of the 14 spherical signatures, namely

∗532	∗432	∗332	∗22N	∗NN
		3∗2	2∗N	N∗
532	432	332	22N	NN

can be obtained from the sphere (for which *char* = 2) by introducing holes, cone points, and corner points — the features symbolized by ∗, **N**, and N, respectively.

These changes to the orbifold decrease *char* by 1, $\frac{N-1}{N}$, and $\frac{N-1}{2N}$, respectively, the costs of these features in the Magic Theorem. An orbifold with these features has Euler characteristic *char* equal to *ch*.

The fourteenth spherical signature is **N**×. The × stands for a *crosscap*, the orbifold obtained by folding each point of the sphere onto the point opposite it, as at right. Bringing together opposite points on the sphere results in a weirdly twisted half sphere, and halves *char*, with *char* = 1.

In Chapter 7 we'll see that each ○ reduces *char* by 2, which is too much to appear in the signature of any spherical symmetry. In Chapter 8 we'll discuss the topology of surfaces, and prove that these are all of the features that an orbifold could possibly have. Thus the list of 14 signatures of spherical symmetry types is complete.

The Magic Theorem for Frieze Patterns

The Magic Theorem for frieze patterns states that the signature of a frieze pattern costs $^\$2$. Its proof is an easy consequence of the one for spherical patterns. This is because we can roll up an infinite frieze pattern into a finite one around the equator of a sphere.

A pattern with signature of the form **NN**.

A pattern with signature ∞∞.

The resulting spherical pattern will have a rotational symmetry of order N, and its symmetry will be one of the seven types ∗22N, 2∗N, 22N, ∗NN, N∗, N×, or NN, whose Euler characteristics have the form $\frac{1}{2N}$, $\frac{1}{2N}$, $\frac{1}{N}$, $\frac{1}{N}$, $\frac{1}{N}$, $\frac{1}{N}$, or $\frac{2}{N}$, respectively. The frieze pattern will correspondingly be one of ∗22∞, 2∗∞, 22∞, ∗∞∞, ∞∗, ∞×, or ∞∞, whose Euler characteristics (obtained by letting N grow to ∞) are 0.

In fact patterns with these symmetries really do have ∞-fold cone points and kaleidoscopic corners — the limits of N-fold ones. As N increases while keeping the length of each period in the pattern the same, these special points will pull further and further away from the rest of the orbifold and ultimately they will "lie at infinity." At right we show orbifolds for **NN** tending towards the orbifold for ∞∞, an infinite

cylinder whose ends are cone points with angle $\pi/\infty = 0$. Turn to page 144 to see more frieze pattern orbifolds with ∞-fold cone points and kaleidoscopic corners.

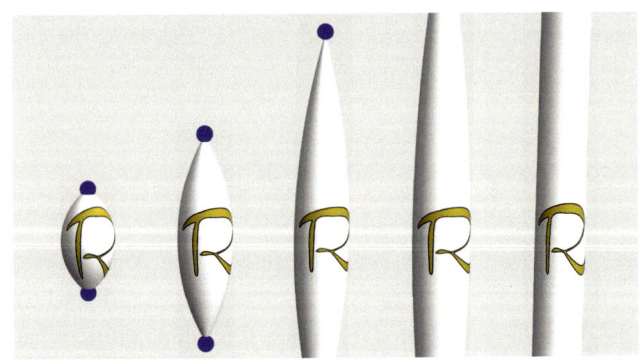

Orbifolds for increasing **NN** *and the limiting* ∞∞.

The Magic Theorem for Euclidean Planar Patterns

Here we must prove that any orbifold corresponding to a pattern in the Euclidean plane has Euler characteristic equal to 0. We do this by showing that, for any really large circular portion of the plane pattern, the Euler characteristic must be close to 0. In the proof we use the fact that the numbers of vertices, edges, and faces inside the circular region are proportional to the area of the circle and so to the square of its radius, while the numbers of vertices, edges, and faces along the boundary of the region are just proportional to the length of the boundary and to the radius of the circle.

To begin the proof, take a map having the same symmetry as the pattern and delete everything that lies outside a circle of large radius R on it. Wrap the circular patch P of the map around a large sphere. This turns a region of our planar map into a map on the sphere, as shown above.

The numbers V, E, and F for the portion P of the infinite map will be close to Nv, Ne, and Nf, where v, e, and f are the (possibly fractional) numbers of these things on the orbifold of the original map and where N is the number of copies of this orbifold completely covered by the portion P.

Since the area of P is just πR^2, this number N will be approximately kR^2 for some positive number k.

In fact, the differences $V - Nv$, $E - Ne$, and $F - Nf$ between the actual V, E, and F and their approximations will be bounded by multiples of R. This is because the "ex-

tra" vertices, edges, and faces belong to copies of the fundamental region of the map that lie across the perimeter of P. The perimeter has length $2\pi R$, so the number of copies of the fundamental region that overlap the perimeter is proportional to R.

We can therefore suppose that $|(V - kR^2v) - (E - kR^2e) + (F - kR^2f)| < cR$ for a constant c, and so

$$\frac{1}{kR}(-c + (V - E + F)/R)$$
$$< ch = v - e + f$$
$$< \frac{1}{kR}(c + (V - E + F)/R).$$

Since the bounds of this inequality tend to zero as R tends to infinity, it must be true that $ch = 0$, completing the proof of the Magic Theorem for planar patterns.

Where Are We?

We have just shown that planar patterns always have *char* = *ch* = 0, proving the Magic Theorem for the plane. In the remaining chapters of this book, we classify the topological features a surface may have and work out their costs. You can turn to Chapter 9 to learn how to make your own orbifolds for many of the patterns in this book, and in Chapter 10 we conclude with the broadest form of the Magic Theorem.

Up to now it has been important to distinguish between the red and blue digits in our signatures because they have different costs. After this chapter, we'll feel free to print them in black. It's easy to recover the proper colors if you want them; the symbols that should be blue are just those before the first cross (×) or star (∗).

Examples and Exercises

We may verify that an orbifold of a planar symmetry type has Euler characteristic *char* equal to 0, even without knowing much about its topology, by drawing a map like the ones on these pages and then counting the (possibly fractional) numbers V of kinds of vertices, E of kinds of edges, and F of kinds of faces in the pattern. These are the same as the numbers V of vertices, E of edges, and F of faces on the orbifold, and we may check that $char = V - E + F = 0$.

By looking at the planar map on the previous page, we can see that its orbifold has $V = 1$ (there is one kind of vertex where the triangles and hexagons meet), $E = 2$ (two kinds of edge, that of a triangle, and that of a hexagon), but $F = 1/2 + 1/3 + 1/6$ (half a rhombus, a third of a triangle, and a sixth of a hexagon) for *char* = 0.

The pattern at right has $V = 1$ (a black dot), $E = 3$ (brown, green, and blue edges), but $F = 1 + 1/2 + 1/2$, be-

cause there are 2-fold cone points in the centers of the blue and green faces. Overall, *char* = 0. In Chapter 8 we will show how to work out that this orbifold has the topology of a crosscap ×, and together with its cone points has signature **22**×.

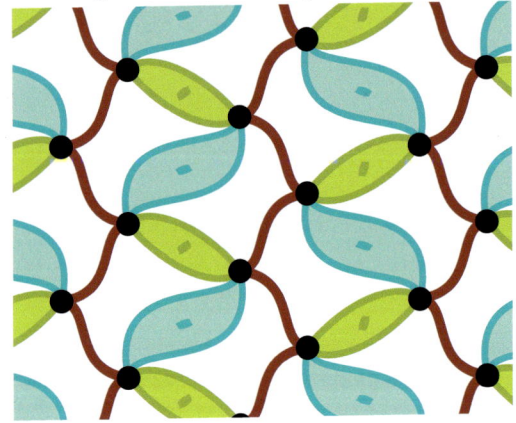

Remember, the numbers of vertices, edges and faces might be fractional! Our answers are on the next page.

QUIZ: *Find the signature of these planar patterns and check that their orbifold Euler characteristic, the sum $V - E + F$ is 0.*

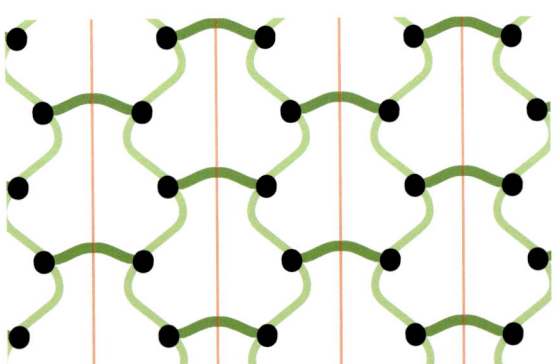

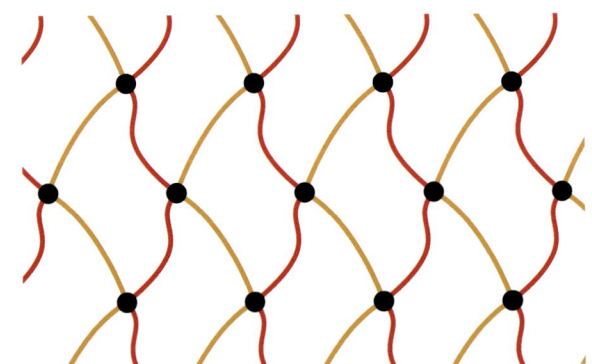

Quiz: *cont'd.* Here are a few more to analyze.

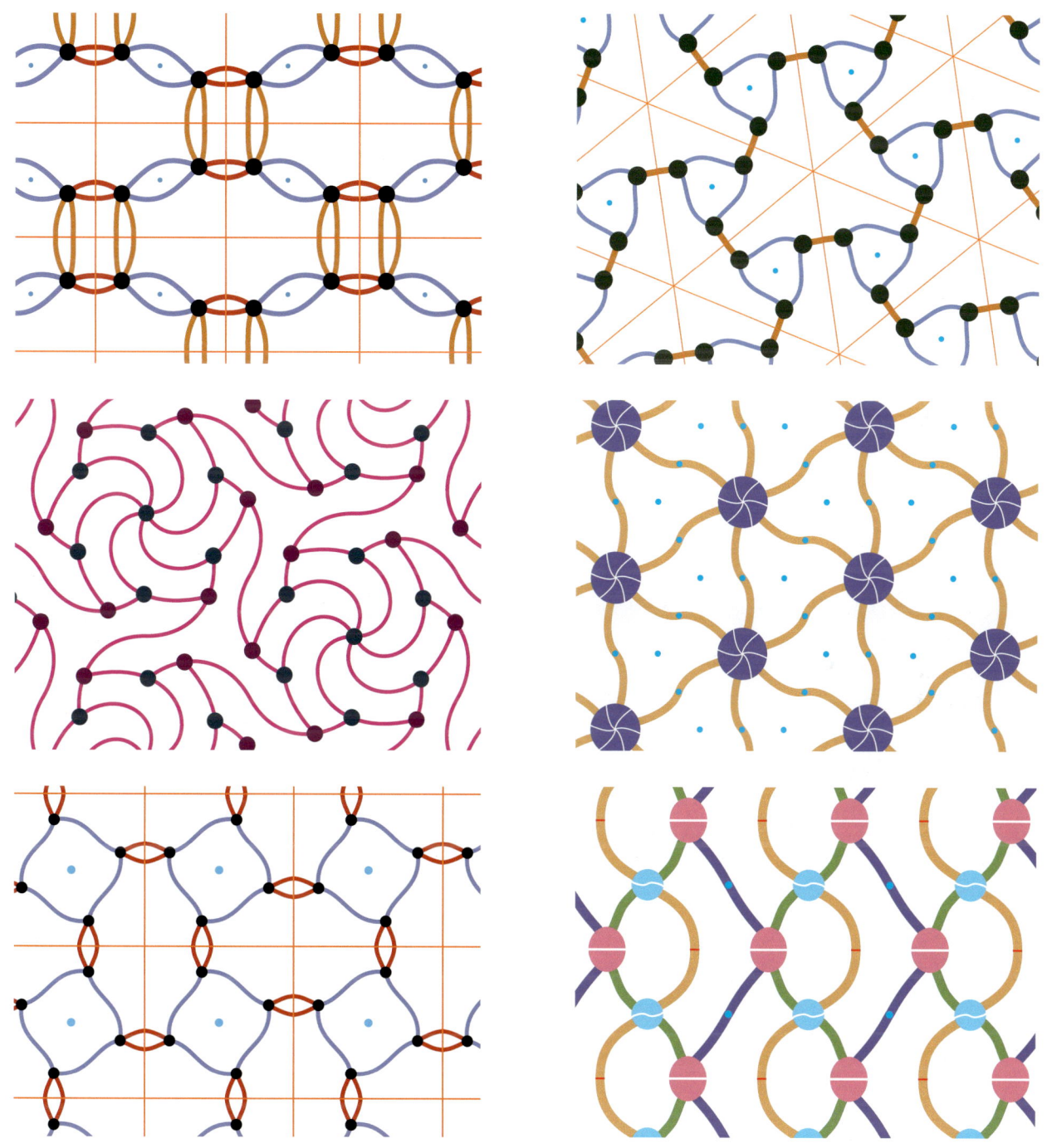

Prev. page, left, $\times*$ with $V = 1$, $E = 1 + \frac{1}{2}$, $F = \frac{1}{2}$; right, $\times\times$ with $V = 1$, $E = 2$, $F = 1$. This page, top left, **2*22** with $V = 1$, $E = 1 + 4(\frac{1}{2})$, $F = 2(\frac{1}{4}) + 3(\frac{1}{2})$; top right, **3*3** with $V = 1$, $E = 1 + \frac{1}{2}$, $F = \frac{1}{6} + \frac{1}{3}$. Middle left, **632** with $V = 2\frac{1}{6}$, $E = 3\frac{1}{2}$, $F = 1\frac{1}{3}$. Bottom left **4*2**, $V = 1$, $E = 1 + 2(\frac{1}{2})$, $F = \frac{1}{2} + 2(\frac{1}{4})$; right **22*** with $V = 2(\frac{1}{2})$, $E = 1 + 2(\frac{1}{2})$, $F = 2(\frac{1}{2})$. All these planar patterns have orbifold Euler characteristic $V - E + F = 0$, as indeed they must. With a symmetry drawing program you can create and check your own!

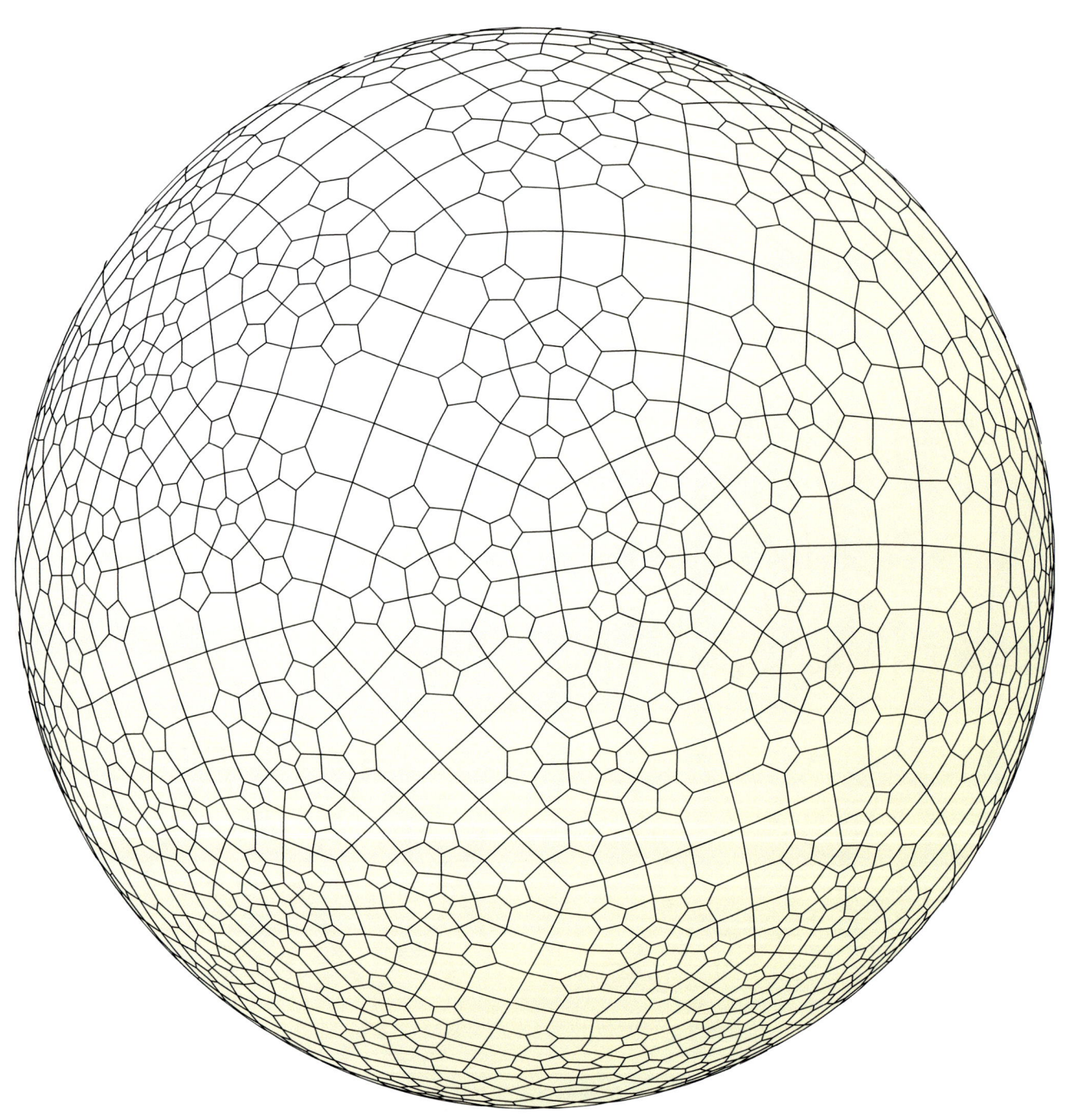

Chapter 7

Euler's Map Theorem

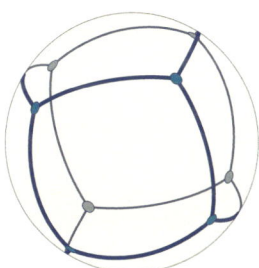 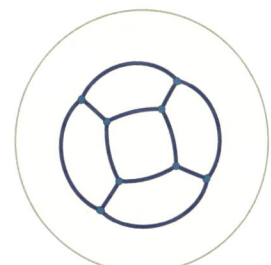

We've made some powerful deductions from Euler's Theorem that $V - E + F = 2$ for maps on the sphere. Now we'll prove it!

Proof of Euler's Theorem

For our convenience, we can copy any map on the sphere into the plane by making one of the faces very big, so that it covers most of the sphere. This is familiar to us: We've all seen maps of the earth drawn in the plane. At the top of the page, we stretch open the back face of a cube, drawing a planar map.

In this planar map, we'll think of this big outer face as the ocean, the vertices as towns (the largest being Rome), the edges as dykes or roads, and ourselves as barbarian sea-raiders!

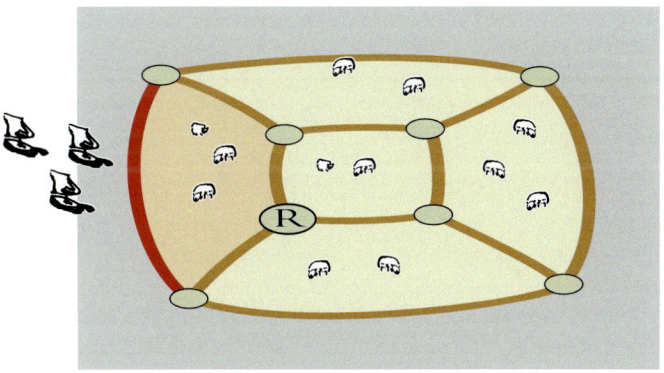

FIGURE 7.1. *Our prey.*

In this new-found role, our first aim is to flood all the faces as efficiently as possible. To do this, we repeatedly break dykes that separate currently dry faces from the water and flood those faces. This removes just $F - 1$ edges, one for each face other than the ocean, by breaking $F - 1$ dykes.

At left, like all maps on the sphere, this beautiful map (signature ***532**) has $V - E + F = 2$.

Deleting an edge decreases the number of edges by 1 and also decreases the number of faces by 1, so $V - E + F$ is unchanged each time we destroy a road and flood a field.

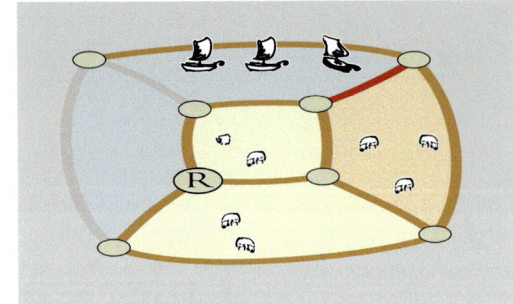

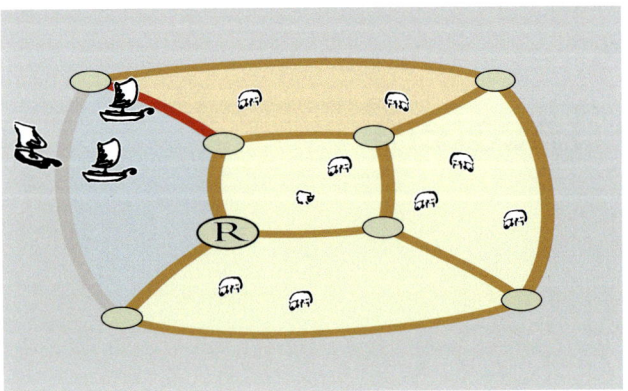

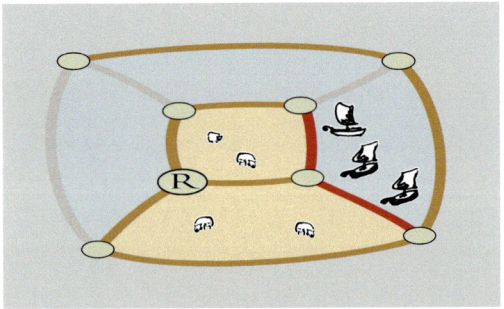

When all the fields have been flooded, we next repeatedly seek out towns other than Rome that are connected to the rest by just one road, sack those towns, and destroy those roads.

(Every town *is* connected back to Rome, because they all were to begin with and we haven't yet destroyed the last road back. There has to be *some* town connected to the rest by just one road, or otherwise a loop of roads would enclose dry fields we have yet to flood.)

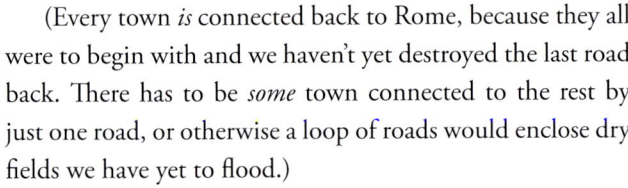

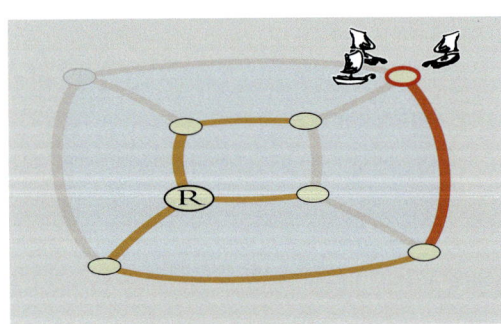

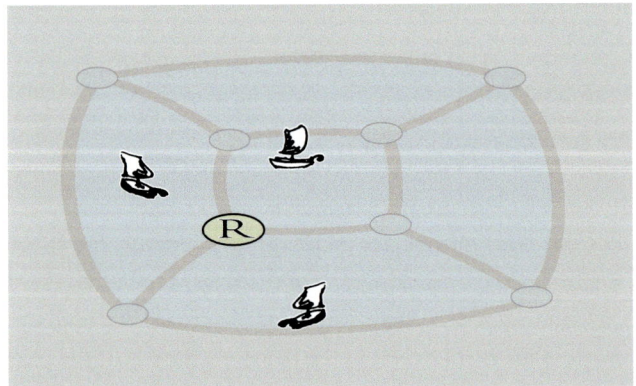

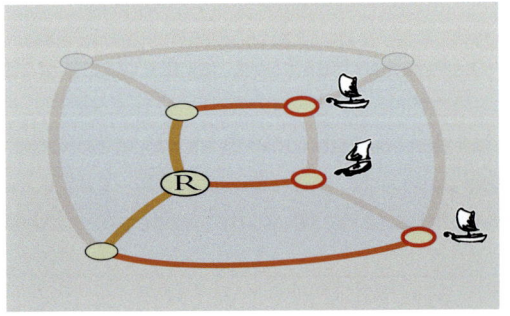

Deleting a vertex and the edge joining it to the remaining roads does not change $V - E + F$ either.

Eventually all of the roads have been deleted, all of the fields flooded, and every vertex but Rome removed.

We have sacked $V - 1$ towns by destroying $V - 1$ roads, one for each town other than Rome. We destroyed $F - 1$ roads to flood all $F - 1$ fields. The number of edges in the original map must therefore have been $(F - 1) + (V - 1) = V + F - 2 = E$. Therefore, $V - E + F = 2$, proving Euler's Theorem.

(Did we sack every town other than Rome? Yes; an unsacked town furthest from Rome would have begun with two paths back to Rome, which however must enclose some dry fields, a contradiction. Did we destroy all remaining roads? Yes; any undestroyed road would be between unsacked towns, which must both be Rome; but this loop encloses dry fields.)

We have tacitly assumed that each face is a topological disk, and we will continue to suppose this. We have also taken for granted some intuitively obvious facts about the topology of the sphere whose formal proofs are surprisingly difficult.

The number 2 is Euler's characteristic number for the sphere. We next show that every surface has such a number.

The Euler Characteristic of a Surface

Theorem 7.1 *Any two maps on the same surface have the same value of $V - E + F$, which is called the* Euler characteristic *for that surface.*

We prove that any two maps on the same surface have the same Euler characteristic $V - E + F$ by considering a larger map obtained by drawing them both together, like these two below are. We shall suppose that no two edges meet more than finitely often, pushing the maps around a bit if necessary.

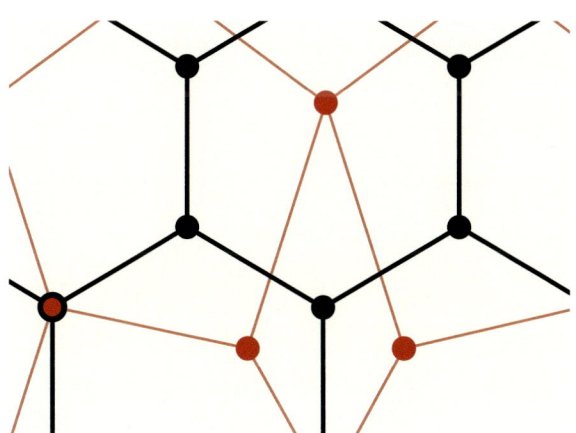

We first draw one map in black ink, the other in red pencil. Then we gradually ink in parts of the pencil map, adding vertices and edges as needed, and noticing that $V - E + F$ does not change.

Inserting a vertex.

V increases by 1, E increases by $2 - 1 = 1$, so $V - E + F$ increases by $1 - 1 + 0 = 0$.

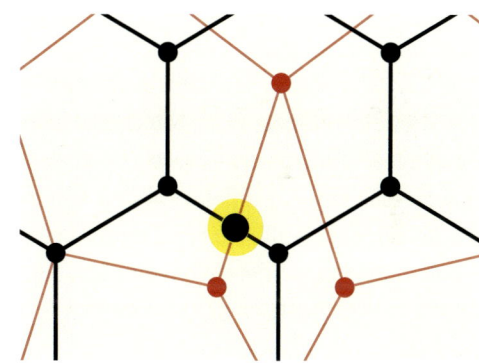

Inserting an edge.

E increases by 1, F increases by $2 - 1 = 1$, so $V - E + F$ increases by $0 - 1 + 1 = 0$.

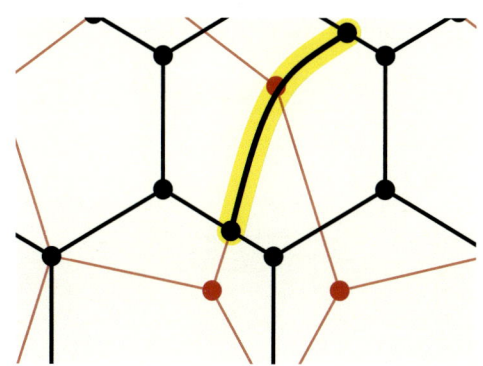

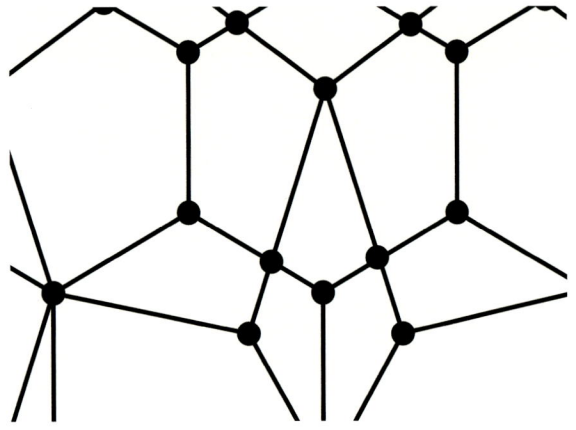

We can continue to make these insertions, gradually inking in the entire figure and not changing $V - E + F$, as at left. Feel free to try this yourself!

This argument shows that the characteristic number $V - E + F$ for the compound map we end up with is the same as that for the originally black map. Equally, it's the same for the originally red map! Therefore, those two original maps must have had the same characteristic, as must any maps on the surface we began with — the Euler characteristic is a constant for each type of topological surface.

The Euler Characteristics of Familiar Surfaces

In Chapter 8 we'll learn more about different types of surfaces. Here we find the Euler Characteristic for a few examples. On each surface, we will obtain the same value for $V - E + F$ regardless of map, so we will choose helpful ones. For maps on more complicated surfaces, we take care to ensure that each each face is a topological disk (which implies the edges and vertices are all connected to one another), and that any boundary is covered by a loop of edges and vertices, counted fully so that Euler's Map Theorem may be correctly applied.

The Euler Characteristic of a Torus is 0.

The map on the torus at left has 16 vertices, 32 edges, and 16 faces, so $V - E + F = 16 - 32 + 16 = 0$.

The map below is much simpler: it has just 1 vertex, 2 edges, and — though this takes some checking — just 1 face, so $V - E + F = 1 - 2 + 1 = 0$. The theorem tells us that we can use either map to work out the characteristic.

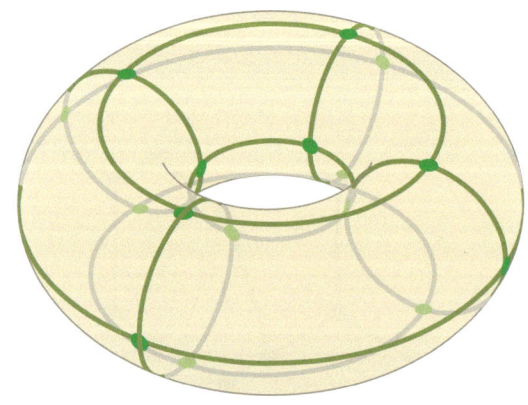

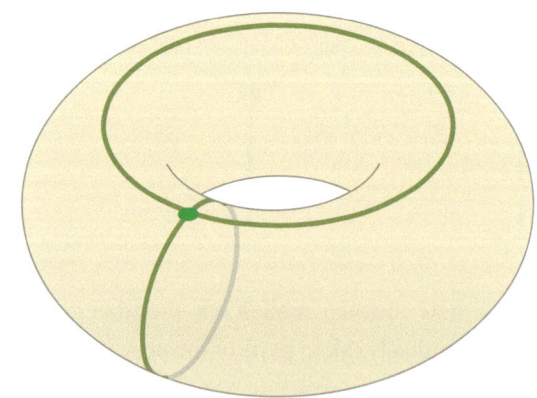

A Disk has Euler Characteristic 1.

A disk always has Euler characteristic 1: the very simplest map has 1 face, 1 edge, and 1 vertex, for $V - E + F = 1 - 1 + 1 = 1$, but any map will do.

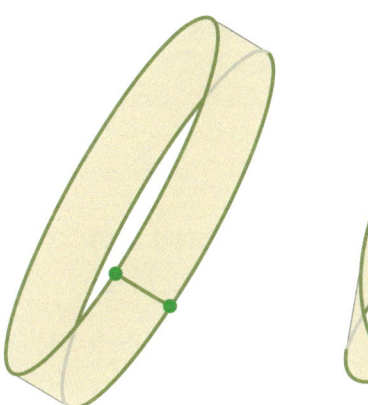

The Euler Characteristic of an Annulus or Möbius Band is 0.

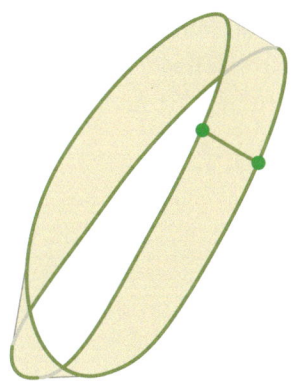

In fact, a disk is topologically a sphere with one hole punched into it. We've met the annulus, with Euler characteristic 0, which may be regarded as a sphere with two (very large) punctures.

On the left, we see a map on an annulus, on the right a map on a Möbius band. Both maps have 2 vertices, 3 edges, and 1 face, and so $V - E + F = 0$. This is so for all maps on a Möbius band or an annulus.

The Klein Bottle Also Has Euler Characteristic 0.

The Klein bottle, a one-sided, boundary-less surface, also has Euler characteristic 0. Again, we choose a map with just 1 vertex and 2 edges, as shown. With a little care you can verify this map has 1 face, and this face is a connected disk, yielding $V - E + F = 1 - 2 + 1 = 0$.

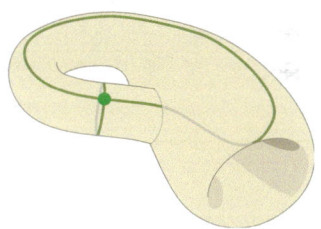

A Sphere with n Holes Punched in It Has Euler Characteristic $2 - n$.

We have already seen that a disk (a sphere with 1 punch) has Euler characteristic 1 and an annulus (a sphere with 2 punches) has Euler characteristic 0. The pattern continues:

We may see this easily by taking a map on the sphere that has a great many more than n faces. If we delete n non-adjacent faces, we have kept V and E the same but decreased F by n.

Consquently, the Euler characteristic will be n less than that of a sphere: $2 - n$. In fact, punching n holes in any surface will always decrease the Euler characteristic by n.

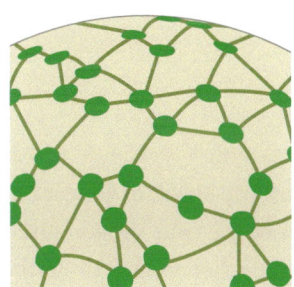
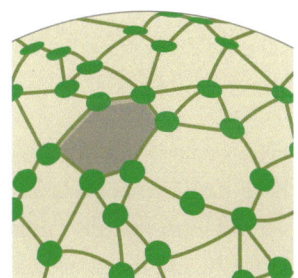

Alternatively, we may systematically design a map specifically for this surface. Here is a specially designed map on a sphere with n punches. It has $2n$ vertices, $3n$ edges, and 2 faces, and so $V - E + F = 2n - 3n + 2 = 2 - n$, and so by Theorem 7.1 must be for any map on this surface.

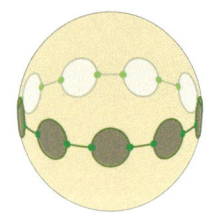

An n-fold Torus Has Euler Characteristic $2 - 2n$.

An n-fold torus is a surface obtained from a sphere by adding n handles, or equivalently n tunnels. We make it by deleting n faces from a sphere and then attaching n handles. Each handle is just a torus with a (very large) hole punched

in it and will contribute $0 - 1$ to the total Euler characteristic. Each hole punched in the sphere will contribute -1. So the net result is that the Euler characteristic of an n-holed torus is $2 - 2n$.

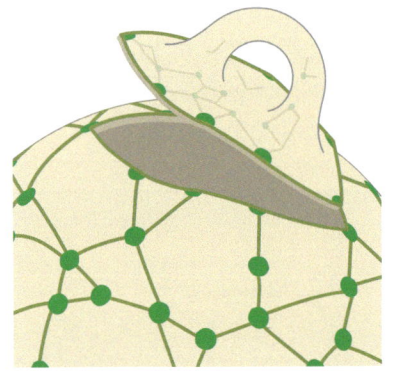

Or, as we show above, we may design a map specifically for this surface with $2n$ vertices, $4n$ edges, and 2 faces: $V - E + F = 2n - 4n + 2 = 2 - 2n$.

Two Mystery Surfaces with Euler Characteristic -2.

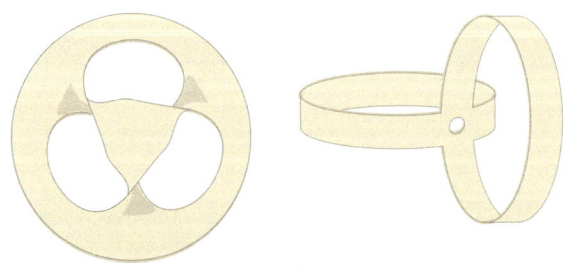

Here we have two mystery surfaces with $V - E + F = -2$. Both have two boundaries and are two-sided; in Chapter 8, we will learn that they then must be the same surface, topologically, and the same as a twice punched torus $\circ **$. In the meantime, you might try to decide for yourself whether this is obvious!

Two More Mystery Surfaces with Euler Char. -2.

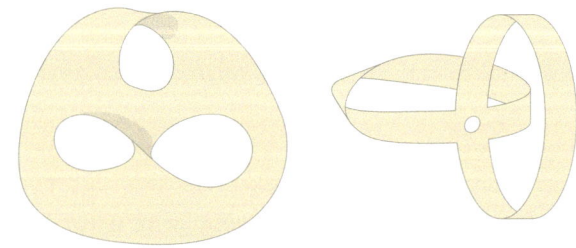

Verify that these surfaces also have Euler characteristic -2. Though they both have two boundaries, they cannot be the same topological type as the two at left, since they are both one-sided, as you should check for yourself. In Chapter 8, you'll find the tools to verify that these are both topologically equivalent to a twice-punched Klein bottle, which we will soon denote $\times \times **$.

In this chapter we have shown that for the sphere the Euler characteristic is 2 and more generally that the value of $V - E + F$ depends only on the surface on which a map is drawn and not on the map itself. This supports the proof of the Magic Theorem in Chapter 6, which in turn supports the enumeration of symmetry types in Chapters 2–5.

In the next chapter we shall classify all possible surfaces, which will show us all the forms an orbifold could possibly take and will help us conclude that we've enumerated the signatures of all possible symmetry types.

There Are Just Five Regular Polyhedra

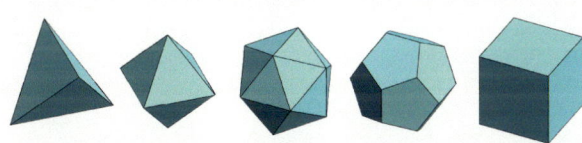

The five regular polyhedra have been known for millennia. Each has one kind of regular polygon for its faces, and the same number of them meeting at each vertex.

Above, from left to right, we see a tetrahedron (regular 3-gons, meeting 3-to-a-vertex, which we will abbreviate $\{3, 3\}$), an octahedron $\{3, 4\}$, an icosahedron $\{3, 5\}$, a dodecahedron $\{5, 3\}$ and a cube $\{4, 3\}$.

Are these all of them? Consider a regular $\{p, q\}$ polyhedron with F p-gons, q of them meeting at each of its V vertices.

Each of the E edges meets 2 vertices, and each vertex meets q edges. Counting all these meetings, we have $2E = qV$. Similarly, each edge bounds 2 faces, and each face is bounded by p edges. Counting these boundings, we have $2E = pF$.

Since $V + F - E = 2$, we find that

$$\frac{2E}{q} + \frac{2E}{p} - E = 2.$$

Since E is positive,

$$\frac{1}{q} + \frac{1}{p} > \frac{1}{2}.$$

Moreover, since the faces of this polygon must have at least three sides, and at least three faces must meet each vertex, $p, q \geq 3$. The only possible values for p and q are on the list above, and so these are all of the regular polyhedra.

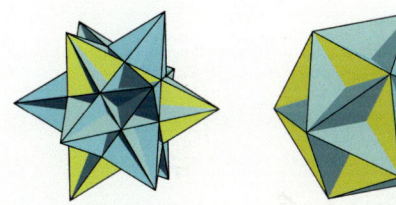

But what do we make of the polyhedra above, with faces that cross through one another? (A face of each is highlighted.)

The great icosahedron at left has 20 triangular faces, five of them meeting edge-to-edge at each of 12 vertices. With 30 edges, it has Euler characteristic 2. Topologically, it is a sphere, partitioned in the same way as a regular icosahedron but tangled up in space in a curious way.

The great dodecahedron at right has 12 regular pentagons as faces, meeting along 30 edges, five faces at each of its 12 vertices. Its Euler characteristic is -6 and topologically, this polyhedron is a four-holed torus. It is not a spherical polyhedron at all!

In the calculation at left we tacitly required a regular polyhedron to be topologically a sphere, and by common sense, we mean an embedded one. The enumeration is complete and there are just the five regular polyhedra at top left.

Chapter 8

The Classification of Surfaces

In Chapters 2–5, we gave a supposedly complete list of symmetry types of repeating patterns on the plane and sphere. Chapters 6–7 justified our method of "counting the cost" of a signature, but we have yet to show that the given signatures are the only possible ones and that the four features we described are the correct features for which to look.

Any repeating pattern can be folded into an orbifold on some surface. So to prove that our list of possible orbifolds is complete, we only have to show that we've considered all possible surfaces.

In this chapter we see that any surface can be obtained from a collection of spheres by punching holes that introduce boundaries (∗) and then adding handles (○) or crosscaps (×). Since all possible surfaces can be described in this way, we can conclude that all possible orbifolds are obtainable by adding corner points to their boundaries and cone points to their interiors. This will include not only the orbifolds for the spherical and Euclidean patterns we have already considered, but also those for patterns in the hyperbolic plane that we shall consider in Chapter 10.

(opposite page) The surfaces shown on these pages, like all other surfaces, are built out of just a few different kinds of pieces — boundaries ∗, handles ○, and crosscaps ×. But it may be hard to tell how, at just a glance! Turn to page 129 to work out the topology of these models crocheted by Shiying Dong.

Caps, Crosscaps, Handles, and Cross-Handles

Surfaces are often described by identifying some edges of simpler ones. We'll speak of zipping up zippers. Mathematically, a zipper ("zip-pair") is a pair of directed edges (these we call *zips*) that we intend to identify. We'll indicate a pair of such edges with matching arrows:

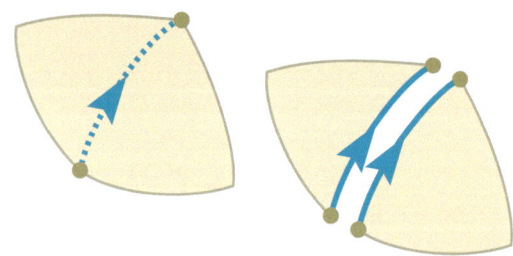

Zipping a Cap

There are simple modifications that you can make to a surface by zipping together the boundaries of one or two holes.

    If a single hole is bounded by a clockwise zip and its counterclockwise mate, we have a *cap*: zipping this up just seals the hole, so we can ignore it.

    Like a sphere or disk, and unlike other surfaces, a cap is *simply connected*: any loop upon it can be contracted to a single point without being hung up on the topology of the surface.

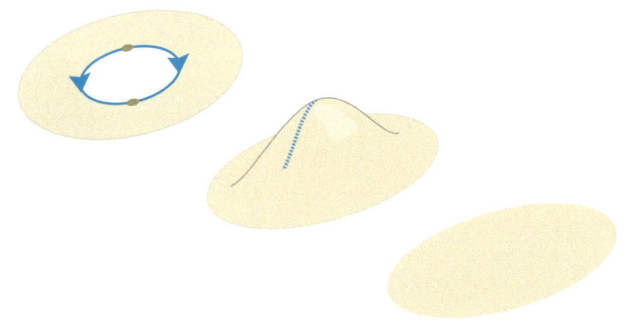

Zipping a Crosscap

If instead the two zips are in the same sense (e.g., both counterclockwise), we have the instructions for what's called a *crosscap*. To get a clear picture is rather difficult: The usual one involves letting the surface cross itself along a line, leading to an 8-shaped cross-section as shown at right.

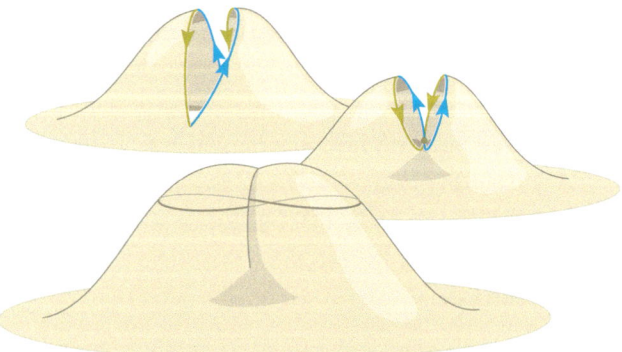

We start by dividing each zip into two zips, as at left. Above we distort the surface, bringing the two sets of zips together. We obtain something like the final surface shown. To really understand crosscaps, draw your own pictures or make some models! (See page 127).

Zipping a Handle

If two nearby holes on a surface are bounded by zips in opposite senses, we have the instructions for a *handle*. To see this, let the two "tubes" grow out of the same side of the surface and then meet, as below.

Zipping a Cross-handle

If such zips are in the same sense, we can let the "tubes" grow out of opposite sides of the surface to form a *cross-handle*, which is sometimes called a *Klein handle*, shown here:

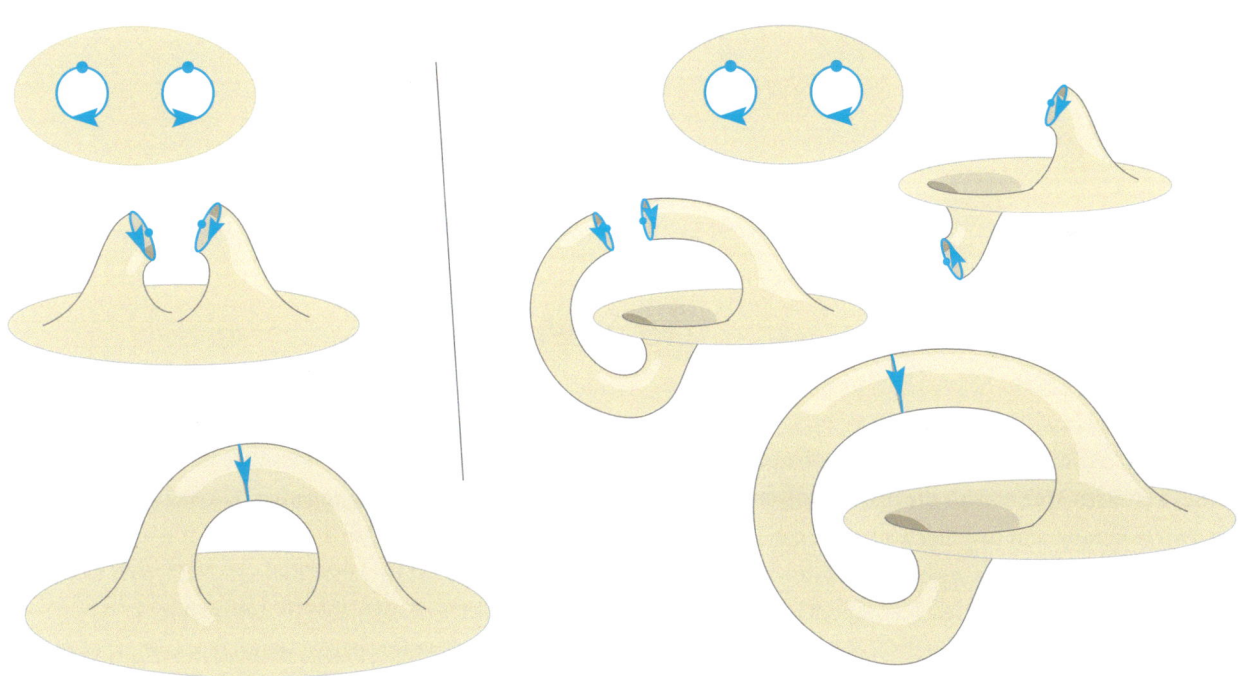

Non-orientability and One-sidedness

Most surfaces we see in ordinary life are *orientable*. No matter what journey the pinwheel makes on a torus, it always returns in the same orientation; the torus is *orientable*.

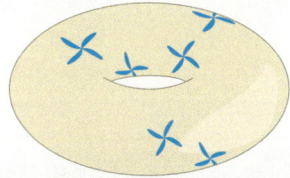

However, like the Möbius band and the Klein bottle (page 127), the crosscap is *non-orientable*, because

if the pinwheel is taken once right round the cap, through the crossing in the middle, it returns in the other orientation — left and right have been swapped.

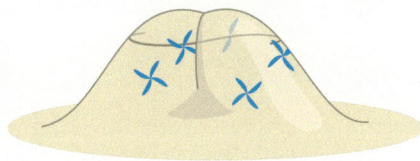

These non-orientable surfaces are also *one-sided* in our surrounding space. It is possible to walk along the surface from each side to the other, and so they are the same side!

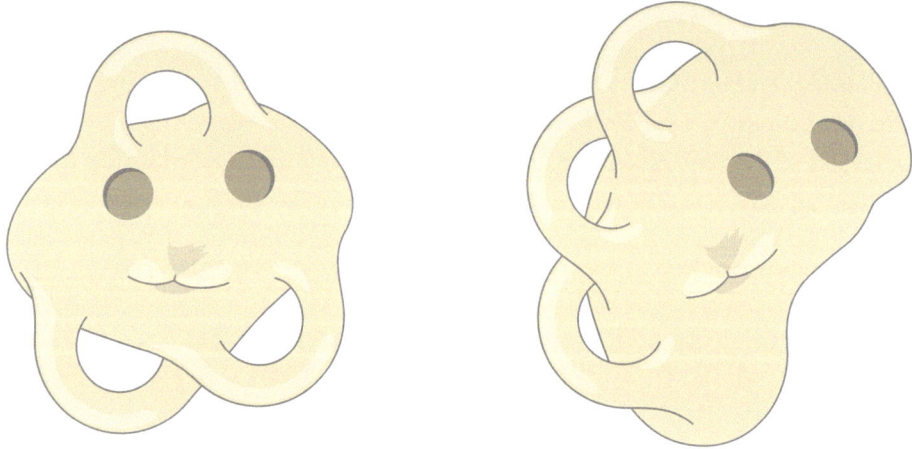

Two equivalent, tidy surfaces. Each has three handles, two holes, and one crosscap.

Tidy Surfaces

In fact, any sphere with the instructions for, say, adding three handles, two crosscaps and two holes is topologically just the same as any other sphere with the instructions for three handles, two crosscaps, and two holes, since we can just push the holes around. The important point is just how many of each of these things there are for each component — each piece — of the surface.

Lemma 8.1 (Tidying Lemma) *Every surface is topologically equivalent to a "tidy" one, obtained from a collection of spheres by adding handles* ○*, holes* *∗*, crosscaps* ×*, and cross-handles* ⊗*.*

To prove this we will suppose that the surface is given to us as a collection of triangles that have zips indicating how they should be pieced together. (In technical language, this is called a "triangulable 2-manifold." It is a deep and difficult theorem, proved by Tibor Rado in 1925, that every compact 2-manifold is triangulable in this manner.)

On the next few pages, we present our proof through diagrams:

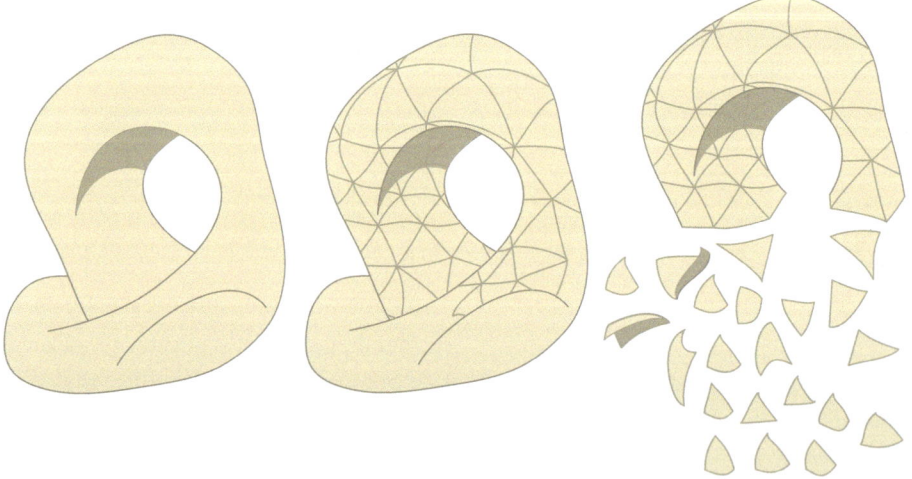

A triangle is already a tidy surface — it's a sphere with a hole in it — and therefore our collection of triangles is certainly tidy before we do any zipping up. So, all we need to prove is that we can zip up any one zip-pair of a tidy surface in such a way as to preserve its tidiness.

The proof is simple in the "snug" cases when the two zips of this zipper together occupy all the boundary components they involve, which we show in Figures 8.1 through 8.3. But Figures 8.4 through 8.8 show that it is almost as obvious in the "gaping" cases when they don't, since these produce the same surfaces as the snug ones, with an extra boundary or two. (The figures illustrate only the "totally gaping" cases.)

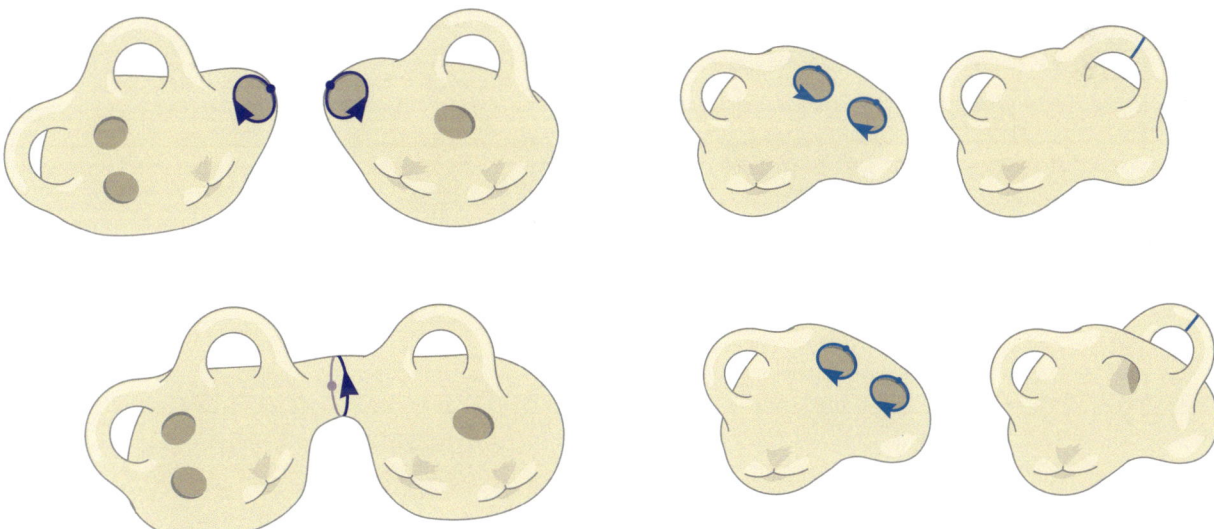

FIGURE 8.1. *Zips on different components of a surface. From* $*^{a+1} \circ^b \times^c \otimes^d$ *and* $*^{A+1} \circ^B \times^C \otimes^D$, *we get* $*^{a+A} \circ^{b+B} \times^{c+C} \otimes^{d+D}$.

FIGURE 8.2. *Zips on different boundaries of the same surface component. At top we zip a pair with opposite orientations. On the bottom we zip a pair with the same orientation. From* $*^{a+2} \circ^b \times^c \otimes^d$, *we get* $*^a \circ^{b+1} \times^c \otimes^d$ *or* $*^a \circ^b \times^c \otimes^{d+1}$ *according to the orientations of the zips.*

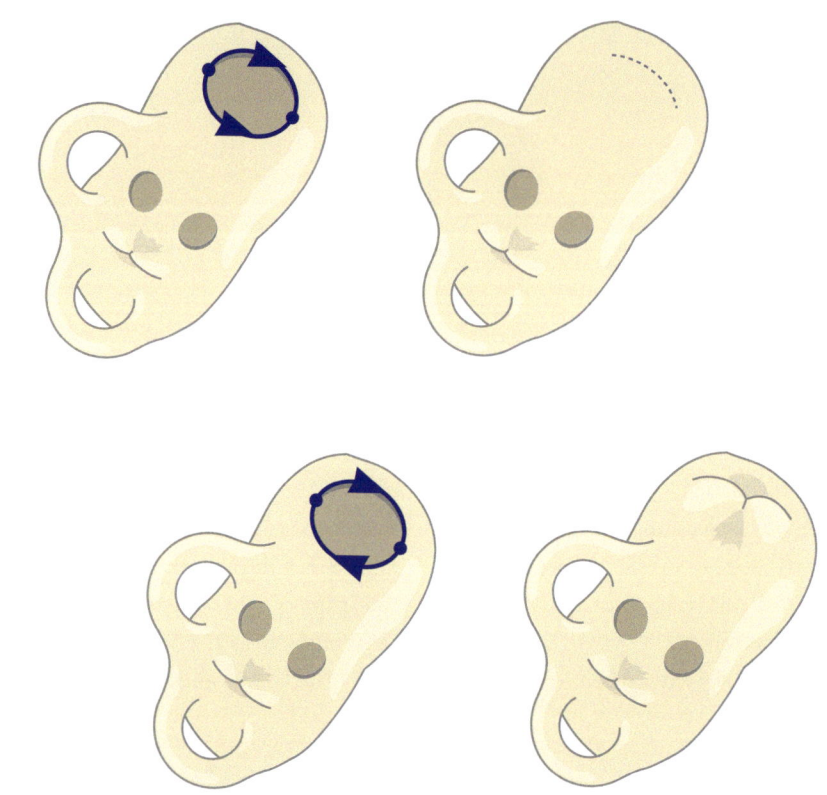

FIGURE 8.3. *Zips on same boundary. From* $*^{a+1}\circ^b\times^c\otimes^d$*, we get* $*^a\circ^b\times^c\otimes^d$ *or* $*^a\circ^b\times^{c+1}\otimes^d$ *according to the orientations of the zips.*

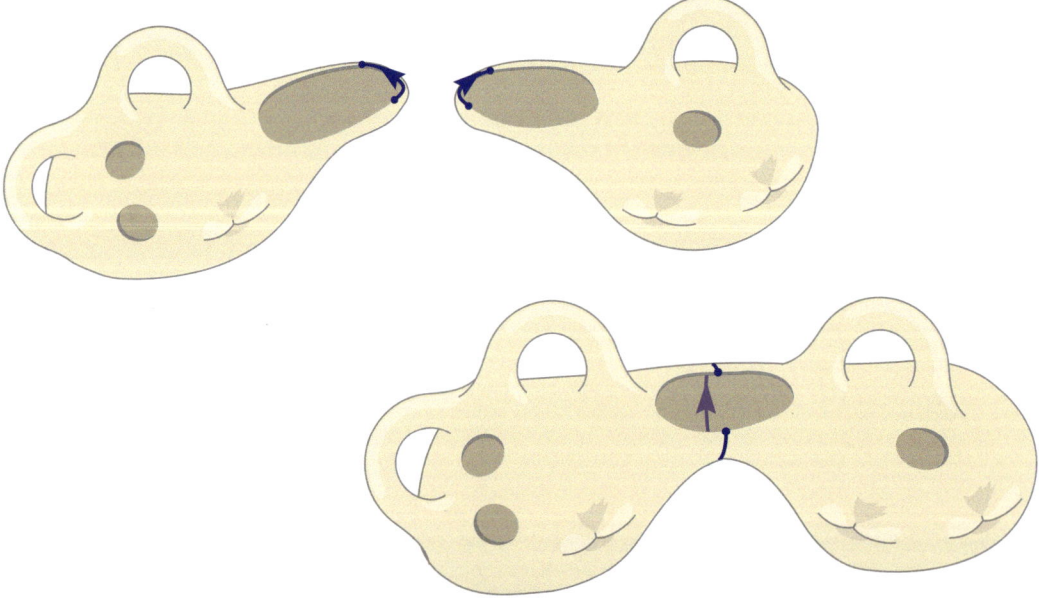

FIGURE 8.4. *Gaping zips on different components form a joined surface with boundary.*

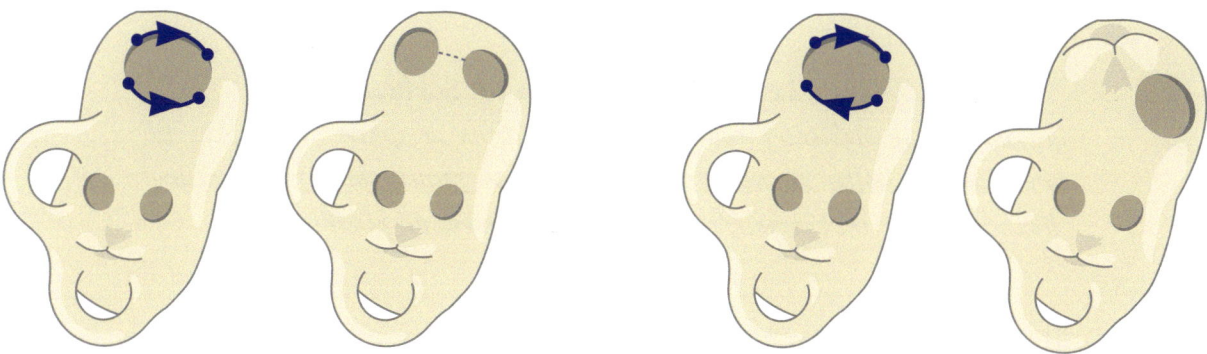

FIGURE 8.5. *Gaping zips with opposite orientations on the same boundary form a cap with boundaries. From* $*^a \circ^b \times^c \otimes^d$, *we obtain* $*^{a+1} \circ^b \times^c \otimes^d$.

FIGURE 8.6. *Gaping zips with the same orientation on the same boundary form a crosscap with boundaries. From* $*^a \circ^b \times^c \otimes^d$, *we obtain* $*^a \circ^b \times^{c+1} \otimes^d$.

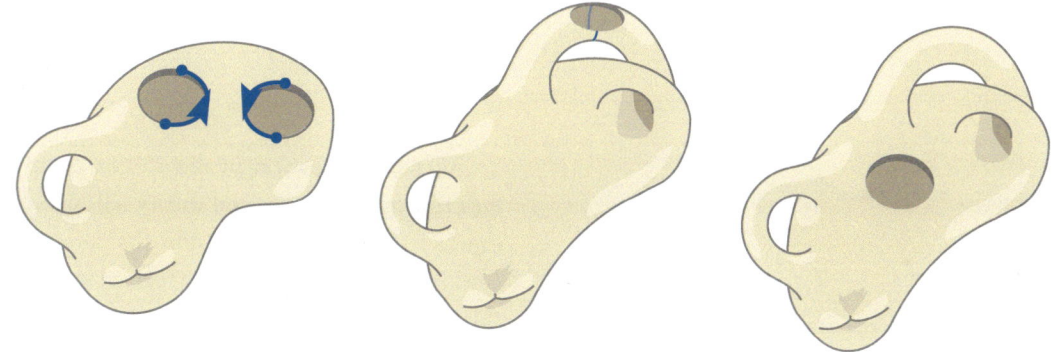

FIGURE 8.7. *Gaping zips with the same orientation on different boundaries of the same component form a crosshandle with a boundary. From* $*^{a+1} \circ^b \times^c \otimes^d$, *we obtain* $*^a \circ^b \times^c \otimes^{d+1}$.

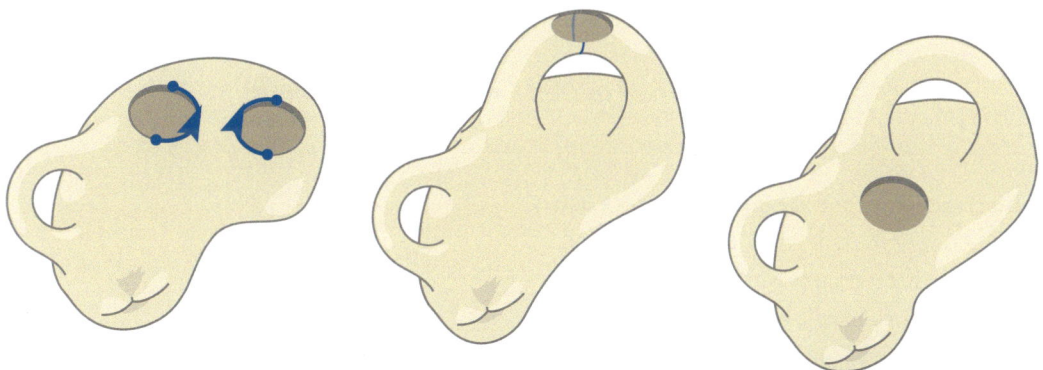

FIGURE 8.8. *Gaping zips with opposite orientations on different boundaries of the same surface form a handle with boundaries. From* $*^{a+1} \circ^b \times^c \otimes^d$, *we obtain* $*^a \circ^{b+1} \times^c \otimes^d$.

In fact we can improve on Lemma 8.1:

Theorem 8.2 (The Classification Theorem for Surfaces)

To obtain an arbitrary connected surface from a sphere, it suffices to add either handles or crosscaps and maybe to punch some holes, giving boundaries. So, the symbols $\circ^a *^b$ and $*^b \times^c$ represent all possible surfaces.

Here's why:

We Don't Need Cross-Handles.

A cross-handle is just a combination of two crosscaps since they are both generated by the same set of instructions. This is shown by zipping up this figure in two ways:

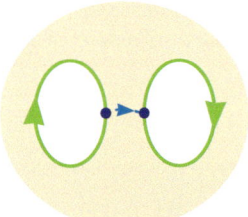

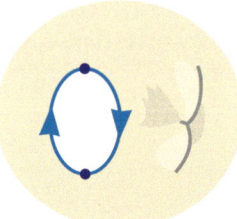

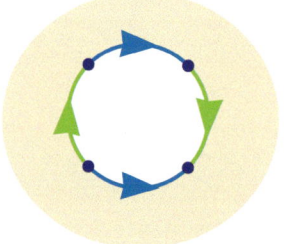

If we do up the blue horizontal zipper first, we get the instructions for a cross-handle ($\otimes$), as in the first figure at top right. Therefore, doing up both zippers will give a cross-handle.

Alternatively, as at right above, if we do up the green vertical zipper first, the general theory tells us we get the crosscap ($\times$) that would come from the corresponding "snug" case, together with a boundary formed by the blue zips. But this boundary is just the instructions for another crosscap, so what we've proved may be expressed by an equation:

$$\otimes = \times \times$$

A cross-handle may be replaced by two cross-caps.

We Don't Need to Mix Crosscaps with Handles.

If we have both a crosscap ($\times$) and the instructions for a handle ($\circ$), we can take one of the holes to be zipped for a "walk" around the crosscap so that it returns with the reversed orientation. The instructions for a handle become the instructions for a cross-handle.

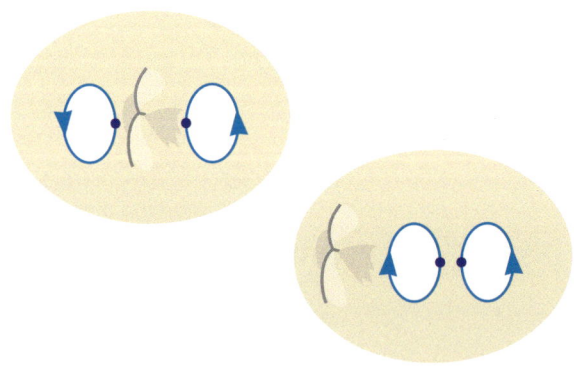

Symbolically, this proves that $\circ \times$ can be replaced by $\times \times \times$ or $\times^3$. More generally, $\circ^a \times^b = \times^{2a+b}$ if $b > 0$.

Euler Characteristics of Standard Surfaces

In the last chapter, we showed that the Euler characteristic of a given surface was independent of its triangulation. To work out the Euler characteristic, we can make our triangulations as nice as we please. The following figures show that

- punching a hole (∗) decreases the Euler characteristic by 1,

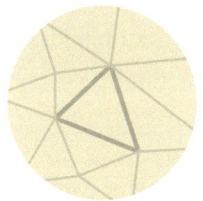 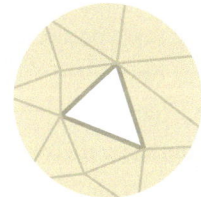

as does

- adding a cross-cap (×),

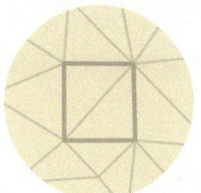 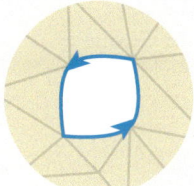

while

- adding a handle (○) decreases the Euler characteristic by 2.

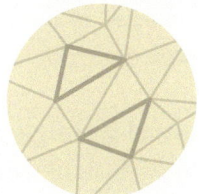 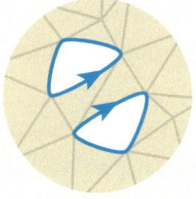

We know that a sphere has Euler characteristic 2; hence, the Euler characteristic is

$$2 - 2a - b \text{ for } \circ^a *^b,$$

$$2 - b - c \text{ for } *^b \times^c.$$

Therefore, to work out the topological type of a surface, count the number n of boundaries of the surface and check to see if the surface is non-orientable, which is the same as having a Möbius band embedded within it. Conversely once we find the Euler characteristic χ of the surface, its type is $\circ^{(2-n-\chi)/2}*^n$ if it is orientable, and $\times^{(2-n-\chi)}*^n$ if it is not.

As a shortcut to calculate the Euler characteristic, as in many of the examples at the end of this chapter, we can slice the surface up, along a (possibly empty) collection of disjoint loops and some arcs between them ("slices"), so that the complement is a topological disk, and then apply this lemma.

Lemma 8.3 (Slicing Lemma) *Suppose S is a surface, other than a sphere and that slices along s arcs and some number of disjoint loops are necessary to slice it into a single topological disk. Then $\chi(S) = 1 - s$.*

Thus if S has n boundary components, its type is $\circ^{\Delta/2}*^n$ if it is orientable, and $\times^{\Delta}*^n$ if not, where $\Delta = 1 + s - n$.

To see why this is so, let S be a surface that is not a sphere. We construct a graph that suits our needs. A surface has a loop on each component of its boundary, or (since S is not a sphere) a non-separating loop if it does not have boundary. On each loop, choose a cycle of vertices and edges — there will always be as many of these vertices as edges, and these will not contribute to our final Euler characteristic count.

Next, we add slices to the graph, paths of edges and vertices, beginning and ending on our initial loops; this may always be done so that the remaining part of the surface can be laid flat as a single, simply connected piece, i.e. a topological disk (If it does not lie flat, there remains something to slice; one never needs to slice the surface into two parts.) Each arc has one more edge than vertex to contribute to the Euler characteristic count, no matter how many there are.

When we are done, there is a single disk as a face, and $F = 1$. No matter how many vertices and edges this graph has, there are exactly s more edges than vertices. Therefore $\chi(S) = F + V - E = 1 - s$. If the surface has type $\circ^{\Delta/2}*^n$ or $\times^{\Delta}*^n$, $\chi(S) = 2 - n - \Delta$, and thus $\Delta = 1 + s - n$.

That's All, Folks!

We cannot simplify this system for describing surfaces any further since all these surfaces are topologically distinct. This is because

- $\bigcirc^a *^b$ is orientable, with b boundary components and Euler characteristic $= 2 - b - 2a$,

while

- $*^b \times^c$ $(c > 0)$ is non-orientable, with b boundary components and Euler characteristic $2 - b - c$,

so that the numbers a, b, and c are invariants.

In particular, we can use one of $\bigcirc^a *^b$ and $*^b \times^c$ to indicate the topological type of an orbifold. But, an orbifold differs from an abstract surface just because it has local features coming from points that were fixed by some symmetries. Since we showed in Chapter 1 that the only possibilities for the symmetries fixing a point are $\mathbf{N}$ and $*\mathbf{N}$, there can be no other local features than gyration points and kaleidoscopic points.

This proves at last that the four fundamental features that make up our signature symbol

wonders	gyrations	miracles	kaleidoscopes
○ ... ○	**A B ... C**	× ... ×	***ab...c *de...f**

really are all that's needed to specify its orbifold. In turn, this finishes our discussion of planar and spherical groups, since we saw in Chapters 3–5 that these are determined up to isotopic reshaping by their orbifolds.

The miracles and wonders in Chapter 2 were just a poor man's way of approaching the global topology of the orbifold surface. We can now formally define them by saying that a pattern "has just a wonders" or "has just c miracles" according as this surface is an orientable one, $\bigcirc^a *^b$, or a non-orientable one, $*^b \times^c$.

Where Are We?

Chapter 1 showed that local symmetries must be kaleidoscopic or gyrational, and in Chapter 2 we added miracles and wonders to obtain our four fundamental features. Supposing that these were enough, we then enumerated the symmetry types of planar and spherical patterns in Chapters 3, 4, and 5, using the Magic Theorem that Chapter 6 deduced from Euler's Theorem, proved in Chapter 7.

In this chapter we have proved the Classification Theorem for Surfaces, which shows that miracles and wonders (now properly defined as crosscaps and handles) can describe the global topology of any orbifold (see also [18]). Putting everything together, this shows that our four fundamental features suffice for the entire structure, so completing the investigation. We conclude that our lists of Euclidean and spherical groups are indeed complete.

Though we've finished our proof of the Magic Theorem, in the next chapter we'll take a look at some of the orbifolds we've been working with, the actual folded and rolled-up surfaces that produce our symmetrical patterns.

Is this all? No! So far we've mentioned only the Euclidean and spherical signatures, which cost at most $2. But we've really classified the more expensive ones too, and we'll see some of the lovely patterns to which they correspond in Chapter 10.

Examples and Exercises

We have shown that every surface is topologically a sphere, possibly with some number of holes, possibly with some number of crosscaps or handles. Our proof of the classification theorem is actually a process for breaking apart our surface and then putting it back together again, in a tidy form. In practice, though, this can be a little tedious.

We can work out the tidy form of a surface just by counting the number of its boundaries, checking whether it is orientable, and calculating its Euler characteristic (which we'll call χ). We can work out χ by drawing a map of the surface or applying the Slicing Lemma 8.3: If a surface requires s slices to be simply connected, then its Euler characteristic χ is $1 - s$.

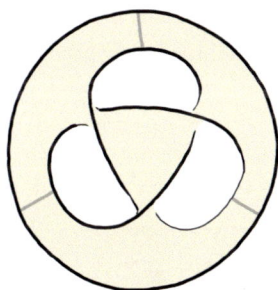

For example, three slices suffice for the strange surface at left to lie flat in one connected piece. Thus it has Euler characteristic $\chi = 1 - 3 = -2$. By tracing around the surface with a finger, we can check that this surface is orientable and that is has two boundaries. The topology of this surface can only be $\bigcirc**$ — this surface is a twice-punched torus! Let's take a closer look and determine the topological types of some other surfaces:

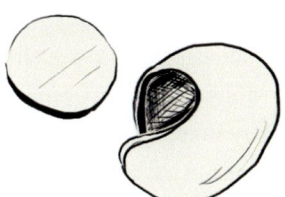

These are orientable surfaces with one boundary. They need no slices to be simply connected, and so have Euler characteristic $\chi = 1$. Therefore each has signature $*$ and is a topological disk.

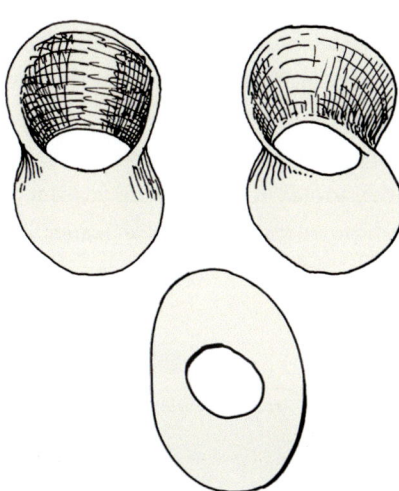

You can check that the surfaces shown at left and below are each orientable and have two boundaries. With a single slice they are simply connected, and so they each have $\chi = 1 - 1 = 0$. Each has type $**$ and is thus a twice-punched sphere, an annulus. We are not concerned with the way that these surfaces are placed into our space — topologically all of these are equivalent.

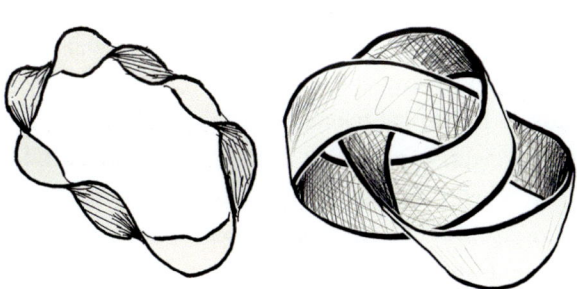

The Möbius Band and Crosscap

What is the topological type of the celebrated Möbius band?

To make your own Möbius band, attach the ends of a strip of paper, putting a half-twist between them, as we've indicated in the figure below. This half twist that renders the Möbius band *non-orientable* (page 117) — taking a clockwise pinwheel around a Möbius band will flip it over and leave it counter-clockwise.

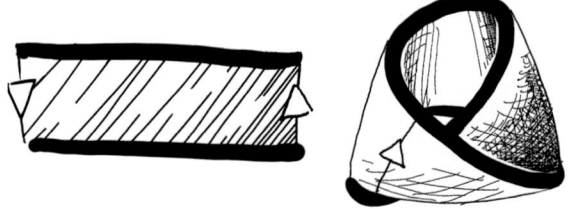

The Möbius band has one boundary, and is therefore a punched form of some surface. In order to work out what that surface is, we can fill in the punch by zipping a disk back onto the boundary of a Möbius band. This is difficult to imagine, but we can build this up in our minds!

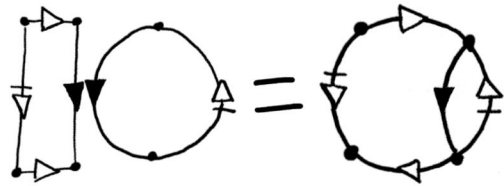

In this diagram, zipping together the top and bottom of the rectangular strip at left yields a Möbius band. If we zip a disk to half of the boundary of this band, we'll have the figure at right. The zips that remain are on opposite sides of a disk. When we zip together opposite sides of a disk we have a crosscap. We conclude that a Möbius band is equivalent to a once-punched crosscap, ×∗.

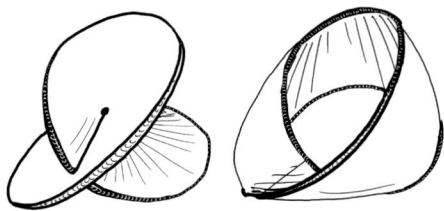

To imagine a crosscap surface, we can wrap a disk around itself as we've shown above left, so that it may be zipped onto the boundary of a Möbius band, producing something like this:

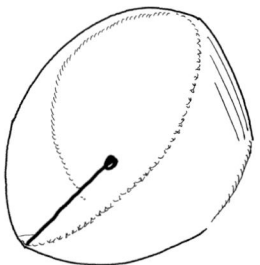

Below we draw another way to visualize the crosscap. In the bottom left, a Möbius band has been punched in the back and its new boundary stretched out into a circle, with its original twist in the middle. Above this, a disk has been arranged so that we can see how it can be zipped on, producing the surface at right.

Adding one punch and filling in another does not change topology — the surface below right is also a Möbius band, with its boundary nicely arranged as a circle, but with its interior tangled up!

We can imagine zipping off this boundary with a disk — closing off the surface in the drawing yields a crosscap surface. From above, it really does look like an × as it crosses through itself!

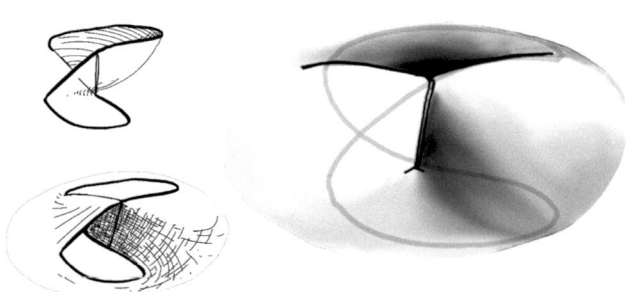

The Torus and Klein Bottle

There are three essentially different ways we can zip together opposite sides of a square.

Zipping together opposite sides of a square without any twists produces a torus ○:

If one zip is twisted and the other is not, we will have a Klein bottle, a cross-handle ⊗ added to a sphere.

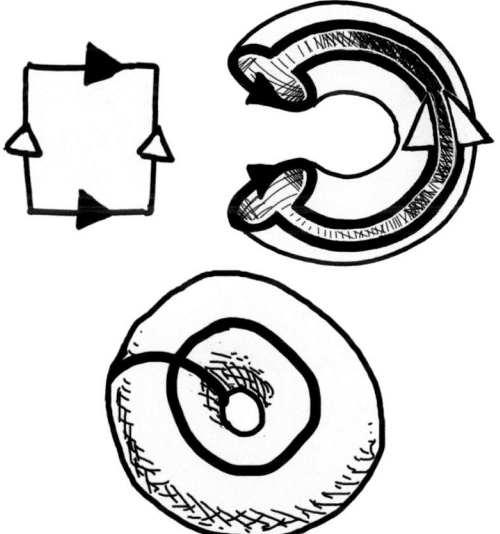

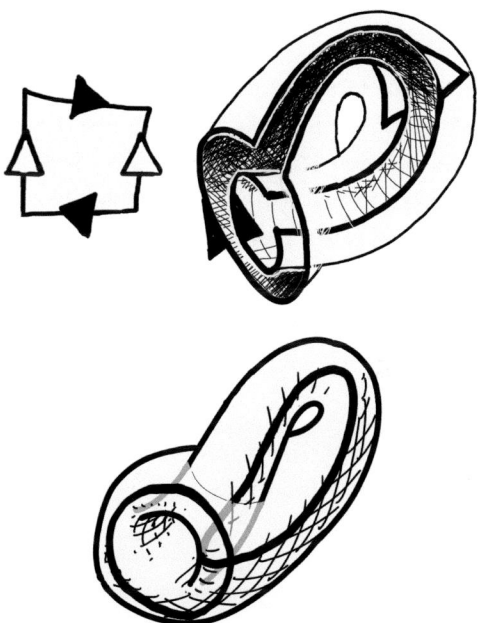

If both pairs of zips are twisted, we are zipping together points that are opposite one another on the boundary of a disk, and the result is a crosscap, something like the drawing below right.

We have learned that a crosshandle surface is topologically equivalent to a sphere with two crosscaps ×× added. Let us try to see this directly by zipping together two punched crosscaps ×∗ along their boundaries:

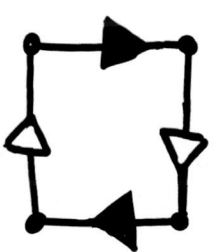

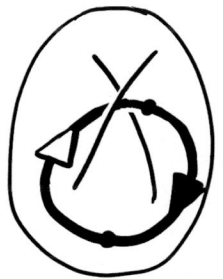

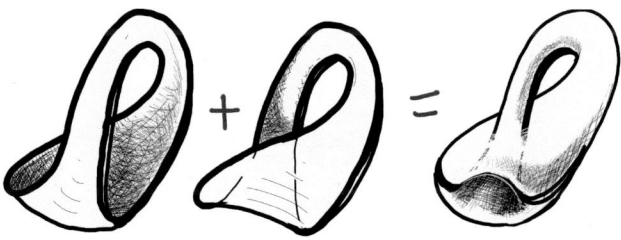

Can you work out how the figures below show Klein bottles?

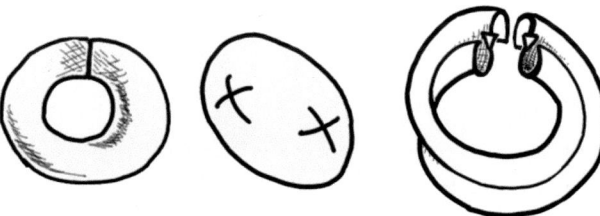

Exercises

1) Look back on page 112 — can you work out for yourself what the mystery surfaces are?

2) Using the Slicing Lemma to find Euler characteristics, what are the tidy forms of these surfaces? Check that both of these strange-looking surfaces are non-orientable, have two boundaries, and have Euler characteristic −1 (two slices render them simply connected). They must both have the same topologicial type. What type is that?

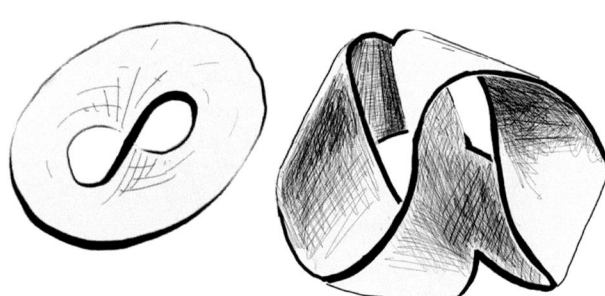

3) The two surfaces below appear to be completely different, but both are twice-punched crosscaps, × × ∗, as you may verify for yourself: They are non-orientable surfaces with one boundary and Euler characteristic −1 (two slices suffice to simply connect them).

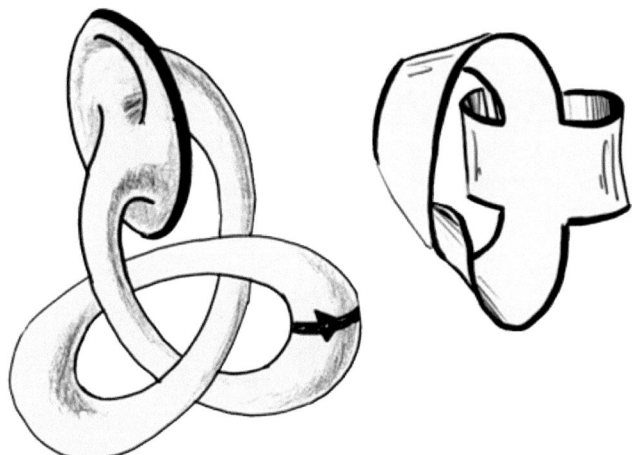

4) Check that each of these surfaces has one boundary, is orientable, and has Euler characteristic −1 (two slices render them simply connected). That is, verify that each of these is a singly-punched torus, ○∗.

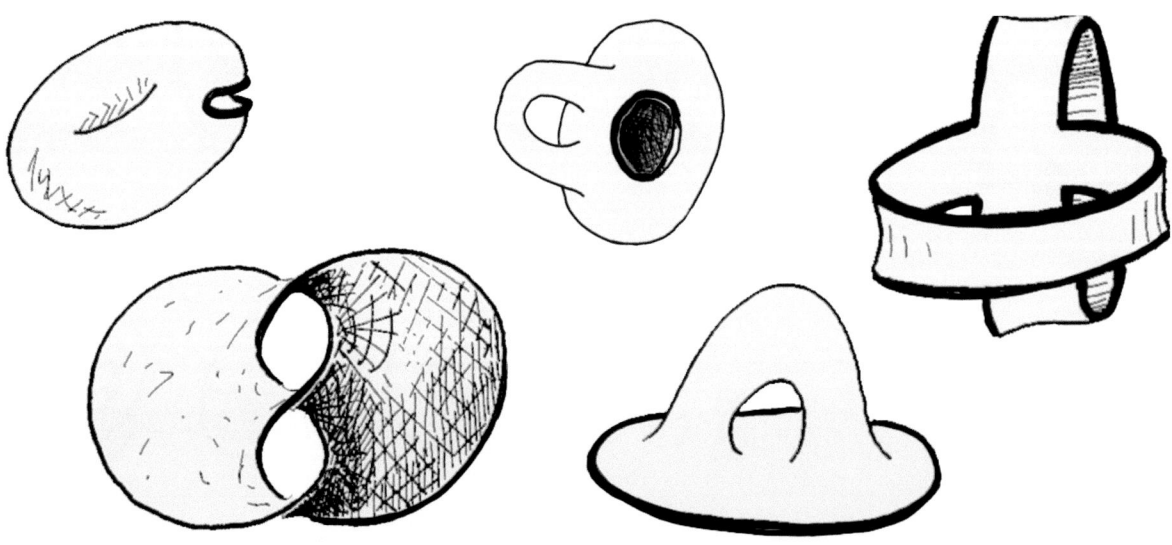

5) You can draw your own mystery surfaces like these to analyze. Draw some loops and shade in their crossings, choosing which alternating regions on the paper will be part of the drawn surface. Check to see if your surface is orientable, and how many boundaries it has. How many slices suffice to render it simply connected? What is its Euler characteristic? — What is the topological type of your surface?

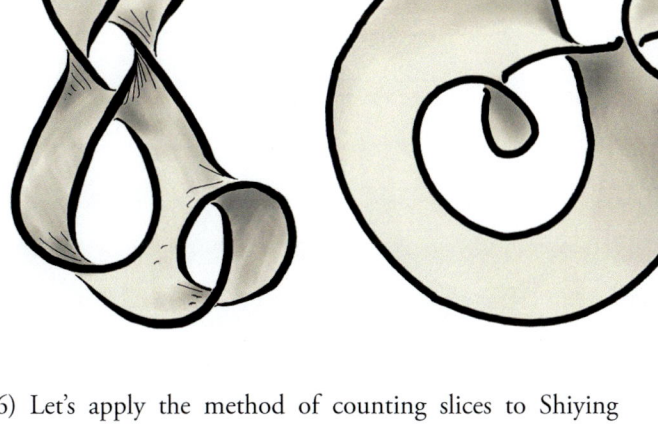

6) Let's apply the method of counting slices to Shiying Dong's crocheted surfaces which we saw at the beginning of this chapter and in Chapter 1. We don't need to see the image on page 115 very clearly to tell that it has the same topology as this surface, which is non-orientable and has two boundaries.

With two slices it can lie flat as one piece; its Euler characteristic is −1 and its type is ×∗∗.

Ignoring its colors, the crocheted surface on page 9 has signature **432**. Its four bounding rings fit nicely on the equators of a cuboctahedron, which we show splayed out

below, to see that five slices suffice. This surface is non-orientable, has Euler characteristic −4, and is therefore ×²∗⁴.

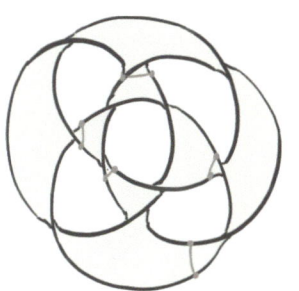

Though the surface on page 114 is fantastically complex, we can use its **532** symmetry type to see that it is assembled from modules with five-, three- and two-fold symmetry. From this we find that $F = 12 + 30 + 20 = 62$. Counting each vertex and edge only once, $V = 240$, $E = 360$, and $\chi = -58$. The surface is non-orientable and has six colored boundaries. Its type is thus $\times^{54}*^6$.

Chapter 9
Orbifolds

The signatures we use to describe symmetry types and prove the Magic Theorem describe features of the orbifold of a repeating pattern. As we've mentioned, we obtain the orbifold of a pattern by equating points that are of the same kind. In this chapter we explore orbifolds in a little more detail — you might recognize some of the patterns we use!

You can read this chapter before, after, or alongside the rest of the book. As you do, we hope you'll keep paper, scissors, and tape on hand so that you can build your own orbifold models for patterns you find on these pages and in the world.

The heart pattern below left has a reflection symmetry across a mirror line. We made this pattern by folding paper along this line, and then cutting a lobe shape into it. (We hope you've all done this before, and if you haven't, please do so right away.) The folded pattern has layers, which we imagine to be fused together into a single sheet to form an orbifold consisting of half of the pattern. This orbifold has a boundary along the fold line in the pattern, and we mark this boundary as *.

The planar pattern on the facing page has many vertical mirror lines. Folding along these mirror lines accordion-style yields a vertical strip with a pattern repeating along it. Rolling that up into a hollow cylinder brings together all of the printed curlicues to coincide as shown at the top of this page. Fusing together points of the same kind produces the orbifold for this pattern — that cylinder shown above — which has the same topology as a sphere with two punches. The signature of this symmetry type is **.

Kaleidoscopic Orbifolds

To make a pattern with kaleidoscopic point symmetry, like this one or the cut-paper snowflakes on page 3, you can fold the paper into wedges that are each $1/N$th of a half circle, all meeting at a point. Cut through all of the layers and unfold your paper to find a snowflake pattern with N-fold kaleidoscopic symmetry and signature *∗N•*.

The orbifold of the snowflake pattern is the single-layered wedge formed by fusing together all of the layers of the folded up paper. Mirrors placed on either side of this orbifold replicate the full pattern of the paper snowflake, as shown below. The boundary of this orbifold, a chain of two mirrors, is what we have been calling a *kaleidoscope*. This boundary is marked as ∗ in the signature of the pattern, and the corner on it as **N**. You may recall from Chapter 1 that we use a • when the symmetries of the pattern fix a point.

The orbifold of the pattern below.

At the start of Chapter 2 we showed you kaleidoscopes for patterns that have signature ∗**632**, like this one does.

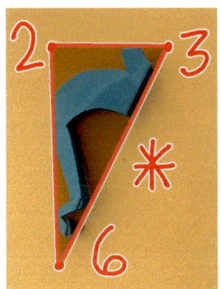

Folding along all of the mirror lines of the pattern at right and fusing together points that are the same kind yeilds the pattern's orbifold — a triangle with corners that are 1/6th, 1/3rd, and 1/2 of a half-circle. To produce the pretty pattern at right, we folded up paper into this triangular shape and cut through the layers.

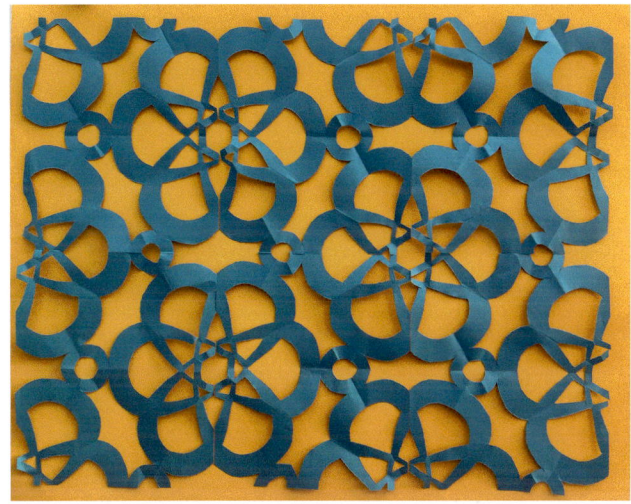

We've seen many "reflecting red" patterns in this book, with planar signatures like ∗**2222**, ∗**332**, or spherical ones like ∗**22**. These patterns have orbifolds that are polygons bounded by mirror lines.

Below and right are cut-paper patterns and their orbifolds, with signatures ∗**442** and ∗**333**, respectively. We made these patterns by folding paper up into triangular shapes. After cutting designs into the paper and unfolding, the foldlines become mirror lines in the pattern. Conversely, we can find the orbifolds of these patterns by folding across their mirror lines and fusing together points that are of the same kind. The boundary of an orbifold (enhanced on the photographs) is formed from these folds, and each piece of this boundary is what we have been calling a kaleidoscope ∗.

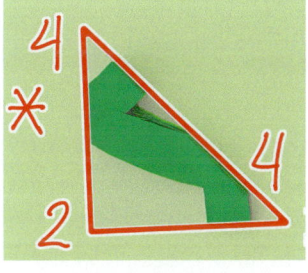

Gyrations and Cone Points

In a gyrational rosette with signature **N**• there are *N* copies of each kind of point around a special center gyration point. If we want to bring together and fuse points of the same kind, there are no mirror lines to fold along as there were in kaleidoscopic patterns. Instead we roll such a pattern into a cone with only $1/N$th of the original material around it. The unique center of this cone-shaped orbifold is a *N*-fold *cone point*. Conversely, if we roll a paper disk into a cone with *N* layers and cut through it, the unrolled pattern will have gyroscopic symmetry like the ones above.

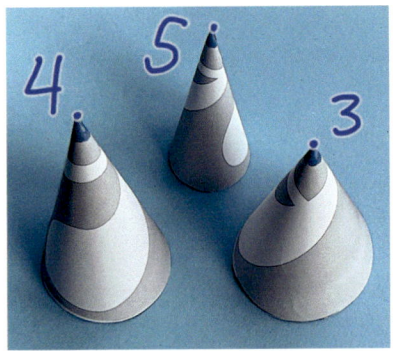

Many crossword puzzles have gyrational point symmetry **2**•, with a 2-fold gyration point at their center. We can split such a pattern into two equal pieces, dividing twin pairs of matching points between them. In the photograph we chose a splitting path in green.

Instead of reassembling a piece with its twin, we can attach its cut edges to each other, forming a cone like the one at right. This cone is the orbifold of the pattern because it has exactly one copy of each kind of point.

All the points on the cone have a complete circle of pattern around them — even the points on the cut are rejoined and made whole — with the exception of the 2-fold cone point at its tip, which has only half of a circle's worth of material around it.

On the other hand, if we have a 2-fold cone, we can split from its cone point along any path we choose, and we will always produce a shape that fits together with a twin copy. The resulting pattern will have a 2-fold gyrational symmetry! This is because each copy has half of a circle's worth of material, and since they meet along curves that were originally the same on the cone, they must fit together perfectly.

Around gyration points orbifolds are cones.

In Chapter 3 we described the "true blue" planar patterns with signatures like **333** or **632**, and in Chapter 4 we encountered the spherical ones. These patterns have gyrations as their only features, and correspondingly the only features of a true blue orbifold are its cone points. For example, a pattern with signature **2222** has an orbifold with four 2-fold cone points. To see this, you can make your own **2222** pattern by splitting open an envelope as its orbifold.

Begin with copies of this rectangle, at least one per person — this activity is best shared with friends!

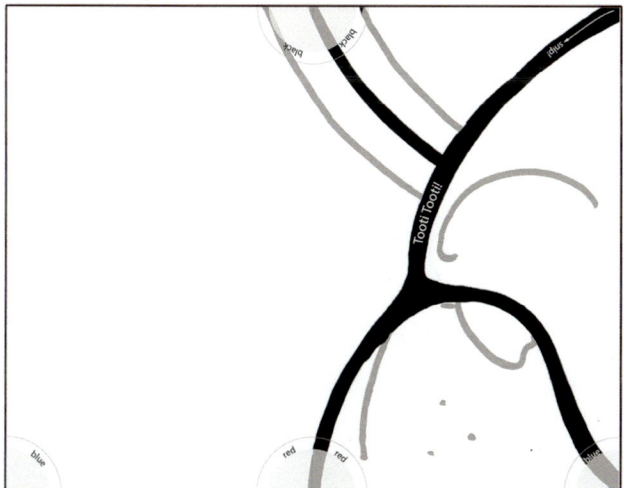

Fold over your paper and tape its sides together to form a kind of envelope. (We've drawn green lines in the photo below to show where to tape.) Next, color the corners of your envelope as shown, everyone in the same way.

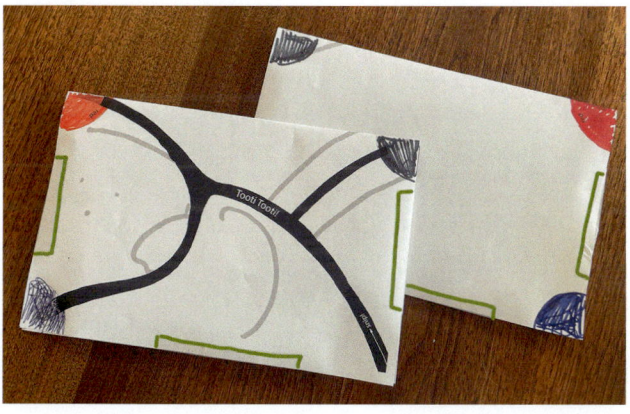

Then cut it open along its printed markings, unfolding it to discover — a dog!

By experimenting, you all will discover that your pieces fit together following a simple rule, that the colors match. The pieces form a regular planar pattern.

This pattern has several kinds of 2-fold gyration points, where the colored markings matched. These points began as the colored corners of the original envelope — the four corners and the four kinds of gyration point are in precise correspondence!

The signature of this pattern is **2222**: there is one **2** for each of the four colors, whether as a corner of the envelope or as a gyration point in the pattern. We named this classroom activity *Tooti! Tooti!* after this signature. Discuss with your friends

How does this trick work?

and turn the page for more!

With scissors and some envelopes, there are infinitely many more tessellating shapes to discover for yourself! So long as your splits reach each corner of an envelope, are connected to one another, and do not divide the envelope into pieces, you will have a tile that tessellates with **2222** symmetry just like the dog shape on the previous page. More than that, the four corners of the envelope will correspond to the four kinds of 2-fold rotation points in the pattern!

It's often surprising what shape you'll find, and it's always fun to decide what it looks like. We split open the square envelope in the photograph above and found a long-necked tessellating chicken:

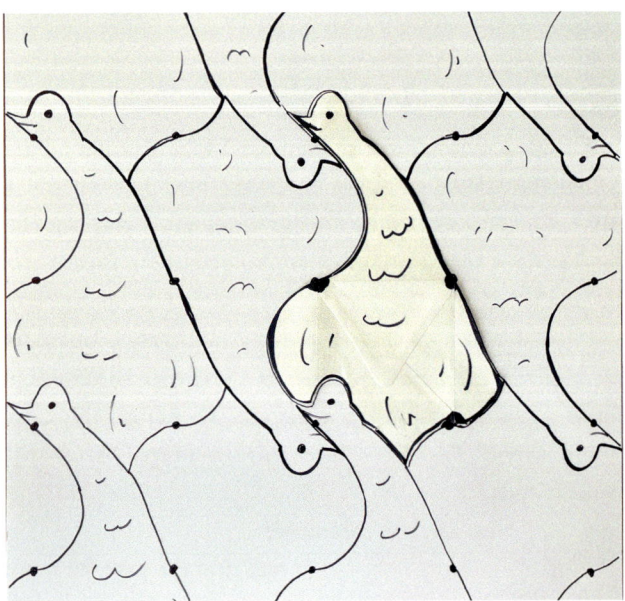

We can understand *why* this shape must tessellate by looking at the four corners of the envelope. On an envelope, most points have their full circle's worth of material about them, but its corners do not — these each have only a quarter of a circle of material on the front and another quarter on the back, for a total of half a circle's worth all together. This means that each of the four corners of the envelope is a 2-fold cone point.

As we saw on at the bottom of page 134, when we split open a 2-fold cone to lay it flat we will always find a shape that fits together with itself to form a pattern with a 2-fold gyrational symmetry, such as at the chicken's shoulder.

In any way that you might split open an envelope to lay it flat, the sides of the resulting shape must match — they matched before they were split apart, and so they match after. Moreover, at each corner there is just the right amount of material at each point for two copies of the shape to fit together perfectly, half a circle's worth each. The shape you'll find *must* be a fundamental region for a tessellated pattern with **2222** symmetry. The four cone points on the orbifold will precisely correspond to the four kinds of gyration point in the pattern. It's fun to try this out!

There are infinitely many ways to split open an envelope, but topologically there are only a few different kinds. Correspondingly, there are just a few different topological types of fundamental region for a pattern with **2222** symmetry — these are called *Heesch types*, which we take up on page 148. Can you find them all in the meantime?

Tessellation Station in the National Museum of Mathematics (MoMath) shows more patterns with gyrational symmetry. This rabbit pattern has signature **333** (if we disregard colors). Can you locate the three kinds of 3-fold gyration point in the pattern?

We can form the orbifold of this pattern by attaching the sides of this rabbit shape to form an equilateral triangle-shaped pillow. Each of its three corners has a third of a circle's worth of material about it — a sixth on the front and sixth on the back. (We've shown two copies of this orbifold to illustrate both of its sides.)

Conversely, this rabbit shape is formed by splitting open the orbifold of type **333**, and any shape made in this way may also tessellate a pattern of type **333**. In such a pattern, there are three kinds of 3-fold gyration point, and these correspond to the three 3-fold cone points at the corners of its triangular pillow-shaped orbifold.

MoMath's Tess the Monkey forms tessellations with **632** symmetry (ignoring colors). Both sides of her triangular orbifold appear in the photograph below, at left. The cone angles at the corners of this envelope are 1/6th, 1/3rd, and 1/2 of a circle, corresponding to 6, 3, and 2-fold gyration points. We've marked these points on Tess — can you find the corresponding gyration points in her tessellation?

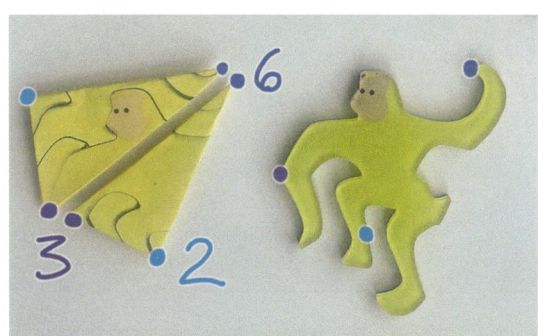

Orbifolds of Hybrid Types

On page 18 in Chapter 2 we met patterns like this one, with signature **3∗3**, that have both kaleidoscopic and gyrational symmetries. Their orbifolds must have both cone points and kaleidoscopic boundaries.

Folding along its mirror lines to bring like points together yields a single triangle in the pattern. Such a triangle is not yet the orbifold because we have not yet brought together all points of the same type — the triangle has gyrational symmetry.

If we cut to its center and bring points of the same kind together, we get the orbifold of signature **3∗3** — a topological disk with one boundary, ∗. A 3-fold cone point is in the interior of this disk and a 3-fold kaleidoscopic point lies on its boundary.

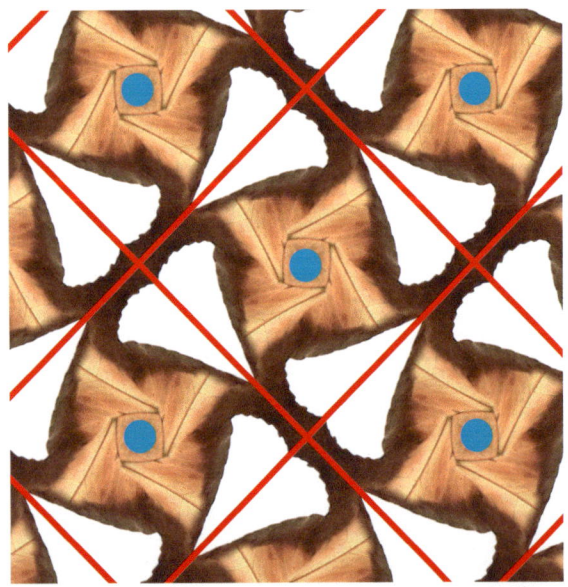

The orbifold for a pattern with signature **4∗2** is also a topological disk with one cone point and one kaleidoscopic point. If we fold up the pattern on the left along all of its mirror lines we will have a single square, but this is not the pattern's orbifold. Bringing like points together forms a cone with an 4-fold cone point at its center.

Spherical patterns with signature **2∗N** or **3∗2** and friezes with signature **2∗∞** have orbifolds with the same topology as the planar ones on this page do — a disk with a marked cone point in its interior and a marked kaleidoscopic point on its boundary. In the hyperbolic plane there are infinitely more patterns with signatures of the form **P∗Q**.

The orbifold of a pattern with signature 2∗22 is a topological disk, with a cone point in its interior, a boundary ∗,
and two kaleidoscopic points **2** and **2** upon it. We can work
out what this looks like, bringing like points together in the
pattern. Folding up the pattern along its mirror lines yields a
rectangle that has two-fold gyrational symmetry. Rolling the
rectangle into a cone to bring together points of the same
kind, we obtain the orbifold shown below.

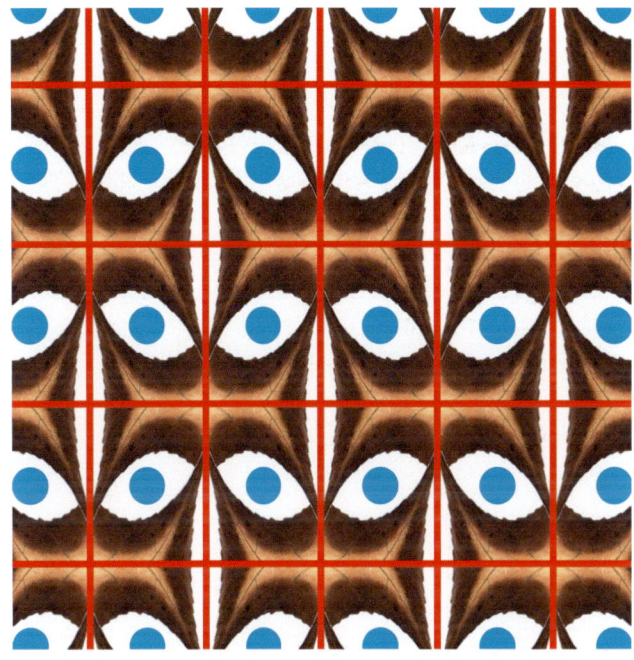

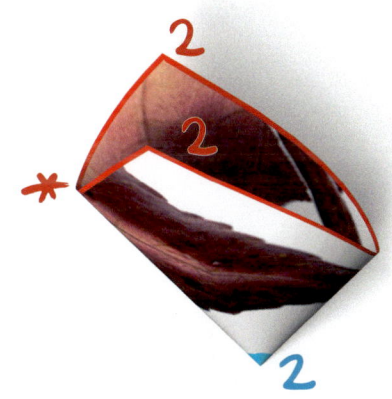

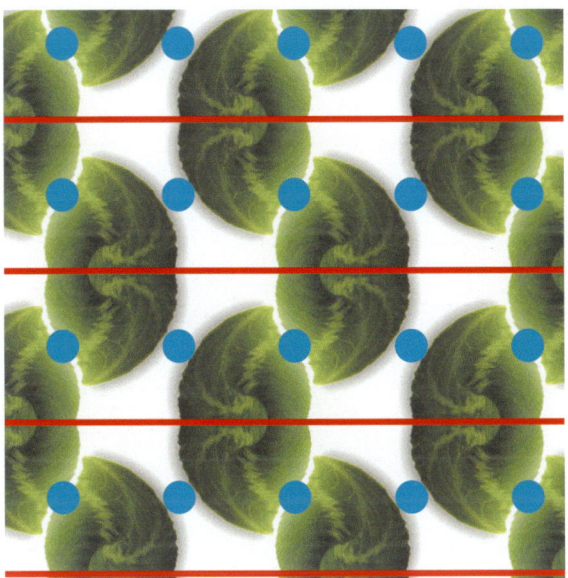

In Chapter 3 we saw this pattern with symmetry type
22∗. If we fuse like points together we will have a pillow-case
shaped orbifold, with two cone points **2**, **2** and a boundary
kaleidoscope ∗, but no kaleidoscopic points.

Tie-dyeing Orbifolds

Carolyn Yackel has folded and rolled fabric into orbifold shapes, which she dyes, producing planar repeating patterns, like these two. You may not be surprised that you can tie-dye a pattern with signature *632 by folding cloth into a triangular pillow. But what is the signature of the lower pattern, and what is its orbifold?

Many symmetry types can't be tie-died, at least not without cutting the fabric — namely those whose orbifolds have cone points or are not embeddable in our space. You can check that there are ten of these. Yackel shows how to dye the other seven in [14].

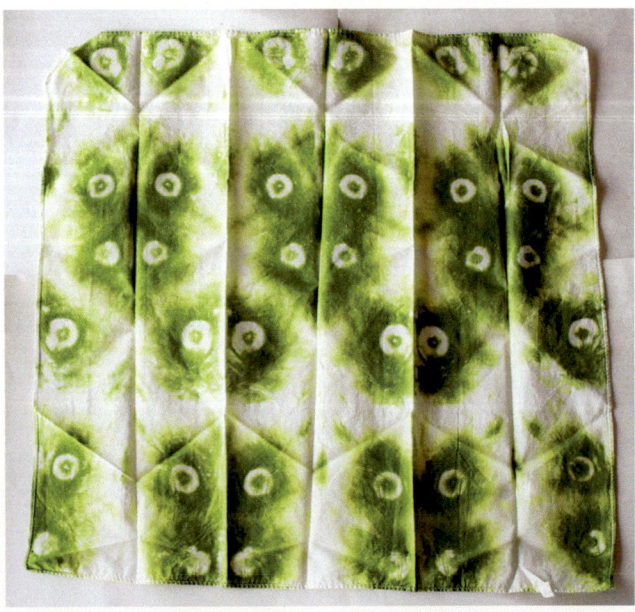

Orbifolds with Topology

So far in this chapter we have focused on "ordinary" patterns, with gyration points, kaleidoscopes, and kaleidoscopic points as their only features. Their orbifolds are simply connected: Topologically they are just spheres or disks, with some marked points. All of the features of these patterns can be recognized by looking directly at them.

As a review, you can match these topological types to the orbifolds and patterns we have seen throughout this book:

Spheres with cone points: signatures **NN**, **22N**, **332**, **333**, **432**, **442**, **532**, **632**, **2222**.

Disks with kaleidoscopic points: signatures *NN, *22N, *332, *333, *432, *442, *532, *632, *2222.

Disks with kaleidoscopic points and cone points: 2*N, 2*22, 3*2, 3*3, 4*2.

Disks with cone points: **N***, **22***.

The remaining symmetry types have orbifolds that are not simply connected. Because of this, these "extraordinary" types can be more difficult to recognize from the patterns directly — we cannot really understand them without examining their orbifolds.

On the opening pages of this chapter, we saw a pattern with signature ∗∗, and its orbifold, a surface with two boundaries. This signature describes the topology of the orbifold and therefore determines the symmetry type.

We now take a look at some of the remaining topological types of the patterns we've encountered.

Möbius Band Orbifolds

Below, a sheet of paper has been folded accordian-style, then rolled up into a Möbius band. Cutting through the layers and spreading out the result produces a pattern with signature ×∗.

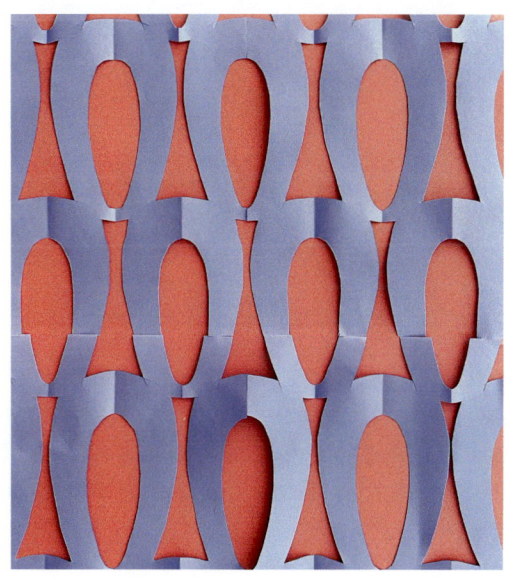

A Paper Torus ○

Here is another amusing paper and scissors trick that you can show your friends: First, glue or tape together two bands, making sure that they are sealed together to form one solid surface, a square with arms linking out of it, like this one:

Now cut down the middle of each of these bands, with the cuts crossing where the bands do. (We've drawn our cutting lines rather boldly.) Unravel result, and you have:

Presto!   A rectangle with a big gap in it?

(Your shape will be missing its middle: we've filled in ours and marked it with a large letter R.) If your cuts are distinctive, as ours are in these photographs, you'll discover that the top of your shape matches the bottom, and the left side matches the right.

Because the sides of your shape match, copies of it can tessellate the plane in a pattern with signature ○.

We can understand this by looking at the orbifold of this pattern of type ○. The left and right sides of a fundamental region match to form a cylinder, and as we showed you on page 127, attaching the top and bottom ends of this cylinder produces a torus — though for a paper model of the orbifold you will first have to mash the cylinder flat, keeping the layers as separate pieces of surface.

Your original cut-apart surface was the rectangle with a gap in it. Therefore, reattaching its sides, our original bands must have formed a torus with a gap in it — a huge gap, the missing square in the middle of the cut open bands.

Klein Bottles

In the same manner, we may attach attach a twisted band to an untwisted one, forming something like the paper model in this photograph.

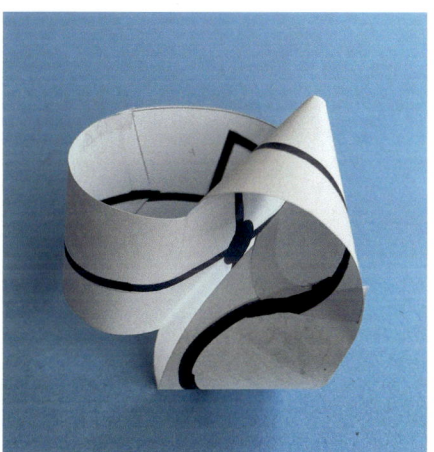

When we cut this open, we again obtain a square with opposite sides that match, but this time a flip is required to match up one of the pairs of sides. We may form a pattern with this square shape and form a pattern like the following

one. Because the orbifold of this pattern is formed by equating points of the same kind, it is the same as what we would obtain by attaching opposite sides of the square shape in this pattern. It has one pair of sides reversed, producing a Klein bottle, ××, as we saw on page 127.

Exercise: In Exercise 2 on page 128, you may verify that a surface made by joining two twisted bands is a crosscap with two holes in it. If you make this surface out of paper and cut it open, you will find that copies of the pieces can tile the sphere with × symmetry!

The Case of **22**×

The orbifold notation gives a unique, well-defined name to each symmetry type, whether in the plane, on the sphere, or even in the hyperbolic plane. This is because the features in a signature record the topology of the orbifold. Other notations — if they are not altogether *ad hoc* — instead record isometries that can generate the symmetries of the pattern, often requiring an arbitrary choice of which generators to use (see Tables A.1 and A.2).

For example, the planar pattern at the beginning of Chapter 3 has signature **22**×. Among its symmetries are two kinds of 2-fold rotation, one kind of glide reflection with a horizontal glide axis, and another kind with a vertical one. The pattern can be generated by the two kinds of glide reflection, or by a glide reflection and a single 2-fold rotation. If we were naming this symmetry type by these isometries, there is no clear choice of which ones we should record.

On the other hand, the topological type **22**× of this pattern's orbifold is fixed and precise: It is a crosscap with two 2-fold cone points, which we may obtain by attaching opposite points together on the boundary of either of these paper models. The two different models show the results of unzipping the orbifold along different kinds of glide axes.

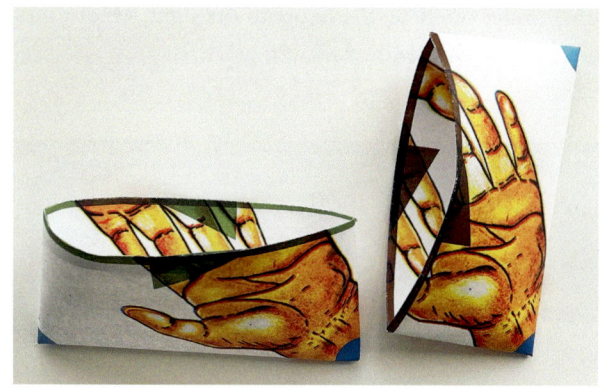

Orbifolds of Friezes

Friezes have just the same sort of orbifolds as other patterns but have kaleidoscopic points or gyration points of infinite order. For example on page 88 we show how a frieze pattern with a kaleidoscopic point of infinite order is the limit of kaleidoscopic point patterns of finite order.

By folding a frieze pattern with signature $*\infty\infty$ accordion-style along its mirror lines, we bring together points of the same kind. We can imagine the orbifold of this frieze pattern as a strip bounded by two mirrors which meet infinitely far away, top and bottom, at an angle of $(180/\infty)°$.

Just as we did with rosettes, we can create a pattern which has this shape as its orbifold. Fold up a strip of paper accordion-style, and then cut through the layers. When you unfold it, you'll have (part of) a frieze pattern with signature $*\infty\infty$.

Similarly, the orbifold of a frieze pattern with signature $\infty\infty$ will be a surface with two infinitely distant cone points of angle $(360/\infty)°$ (page 102). This orbifold is modeled by a rolled up piece of paper! If we cut through all the layers of paper under the R on the figure to the right, we can unroll the paper to get the frieze pattern below.

We can make a paper model of the orbifold of any frieze pattern, recording cone points and kaleidoscopic corners of infinite order, infinitely far away. Counting these points in, the orbifolds of the frieze patterns are topologically the same as their spherical counterparts.

An orbifold of type ∞∞ is an infinite cylinder, a sphere with a pair of cone points of infinite order that we mark ∞. Indeed, for thousands of years people used orbifolds of this kind to roll out a frieze pattern as their signature! We show such a cylinder seal on page 7.

An orbifold of type *∞∞ is an infinite strip, the two sides meeting infinitely far away at a pair of kaleidoscopic corners of infinite order on its boundary. This is topologically just a disk, or punched sphere, with two points marked on its boundary as these corners.

An orbifold of type **22**∞ is a topological sphere with three marked cone points, two that are 2-fold and one of infinite order. Geometrically, it is an infinitely long rectangular pillowcase.

An orbifold of type *22∞ is a topological disk with a boundary *. On this boundary there are three kaleidoscopic points, two of order 2 and one of infinite order. The orbifold is one end of an infinitely long strip.

An orbifold of type **2***∞ is a topological disk with a cone point of order 2 in its interior and a kaleidoscopic point of infinite order on its boundary. It looks something like the illustration at right.

An orbifold of type ∞* is a topological disk with a cone point of infinite order in its interior. Geometrically, it is an infinite cylinder, capped at one end by a boundary * and with the cone point of infinite order at the far end.

Finally an orbifold of type ∞× is a topological crosscap with a cone point of infinite order. We can visualize this as an infinite strip with opposite sides attached, or as something like the orbifold of ∞* but with opposite points on its boundary fused together.

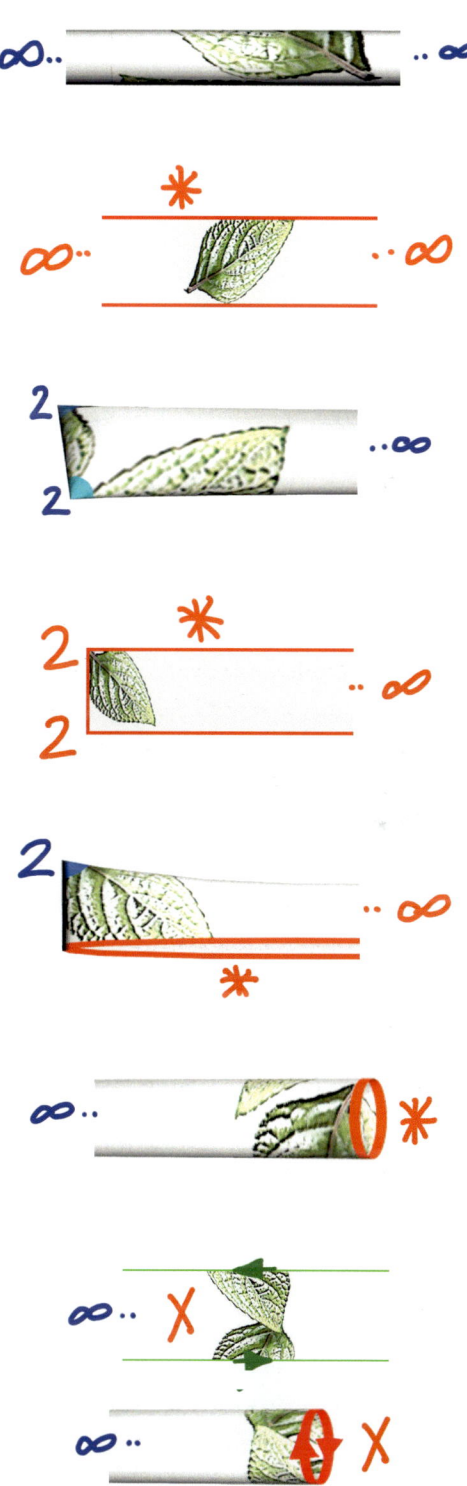

Where Are We?

In this chapter, we've shown you many examples of orbifolds for the kinds patterns in this book. In the next chapter, we'll take up the general case, stating the complete form of the Magic Theorem. We'll close this chapter with a few applications and observations.

A puzzle:

On page 88 we saw a hall of mirrors trick, formed by reflections between two parallel mirrors. In a similar manner, you could trap a beam of light reflecting back and forth between them forever, at least if you could ever get the source of the light out of the way! Our question here is:

Can you trap a beam of light between two mirrors that are not parallel?

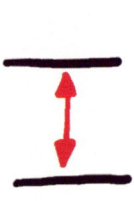   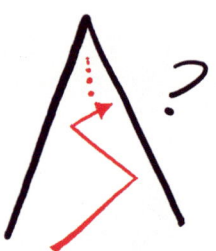

Is it possible, as in the figure above right, for a beam of light to enter into a kaleidoscopic chamber, yet bounce forever and never re-emerge? A little experimentation and intuition suggest not, but we would like an explanation.

Perhaps you have stood in front of a mirror and imagined that a duplicated copy of our world really lies beyond it. If you shine a flashlight into a mirror, it will look as if its beam has passed straight into this duplicated mirrored world! This is because the angle of incidence is the same as the angle of reflection:

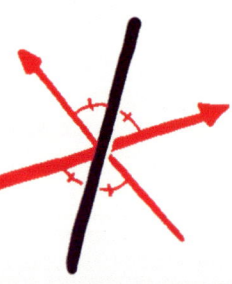

In the same way, we can imagine that what we see when we look into a chamber between two mirrors really does exist just as we see it, with many duplicates of the region between them. The real chamber is the orbifold of the pattern as it appears to us. In the photograph below, two mirrors meet together to bound an orbifold with signature *4•.

You can see for yourself that when we shine a beam of light into the chamber it appears to cross straight into the space that we see and can only cross a few mirror lines before it disappears. Correspondingly, a beam of light bounces finitely many times before exiting the real chamber.

In order for a pair of mirrors actually to bound an orbifold of a kaleidoscopic pattern, they must meet at a special angle, an even integer fraction of the circle. But for any angle, a beam of light will appear to continue in a straight line, and eventually must exit the reflected chambers that we see. Light will always escape.

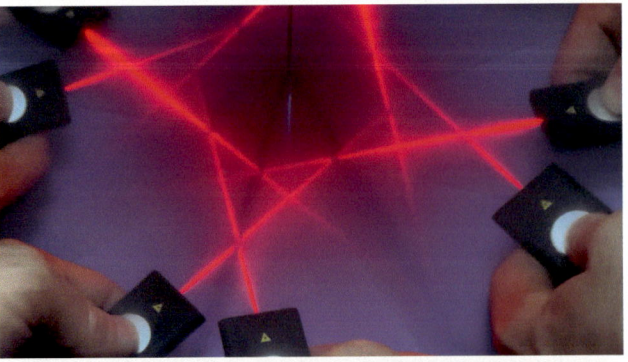

Isotoping Orbifolds

On page 43 we saw that some types of pattern may be isotoped continuously without changing their symmetry type. For these types, we may correspondingly isotope their orbifolds, changing their geometry, but leaving their topology unchanged.

Patterns with signature *2222, 2*22, 22*, and 22×, as well as *× and **, have orbifolds that are fundamentally rectangular, and we may change the ratio of this rectangle's height to its width. Correspondingly there is a single choice in the geometry of these orbifolds, and in the pattern themselves, up to scale or positioning.

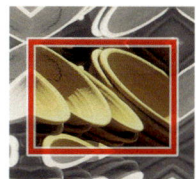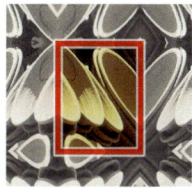

More interestingly, we have two degrees of freedom when we isotope patterns with signature ○, ××, or 2222. Let's take a look at the latter.

The 2-fold rotation points in a pattern with signature 2222 will lie on a lattice, and this lattice may be continuously isotoped, by changing the direction vectors that determine it.

If the 2-fold gyration points remain on a rectangular grid, the shape of the orbifold remains a flat pillow: the envelope that we first encountered on page 135.

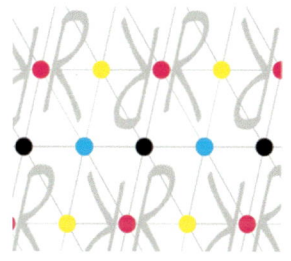

We can shear a pattern of this signature so that its gyration points lie on a skew lattice, as above. The resulting orbifold will be a tetrahedron. Its four faces will be copies of an acute triangle, as shown in the pattern above. At each vertex of this tetrahedral orbifold, we will find one copy of each of the triangle's three corners. These together form a half-circle's worth of material, and the corners of this tetrahedron are therefore 2-fold cone points.

At a special moment of this isotopy, the gyration points lie on a lattice of equilateral triangles as in the pattern below. At right is a pretty computer-generated image of its orbifold, puffed out to better show its cone points. A paper model of this will be a regular tetrahedron with flat faces.

Heesch Types

In how many topologically distinct ways can we choose a shape to be a connected fundamental region for any of the symmetry types? This subject was investigated by Heesch and independently by M.C. Escher [11]. Branko Grünbaum and G.C. Shephard give an exhaustive enumeration of the plane types in [8]. Daniel Huson and Olaf Delgado pioneered a much simpler and more general treatment using orbifolds, which we sketch here and treat more fully in Chapter 16 of *The Symmetries of Things*.

The answer, of course, depends on the symmetry type — obviously for a reflection group the fundamental region is unique. On the other hand, there are four topologically different kinds of fundamental region for **632**. Why is this? The answer is found by looking at the orbifold: A graph on the orbifold will be the boundary of a fundamental region if it cuts the orbifold into a topological disk which has no internal cone point, and can be opened flat onto the plane — a tile. We cut along a graph like this when we split open envelopes to make tessellations on pages 135 and 137.

There are two topologically different ways to split open a sphere that has three cone points into such a tile:

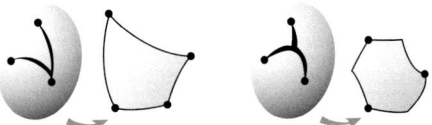

If the cone points have distinct orders, as for **632**, we get four possible graphs altogether: There is essentially only one way to label the cone points of the second graph, but the other can be labeled in three different ways. So there are four topological kinds of fundamental region.

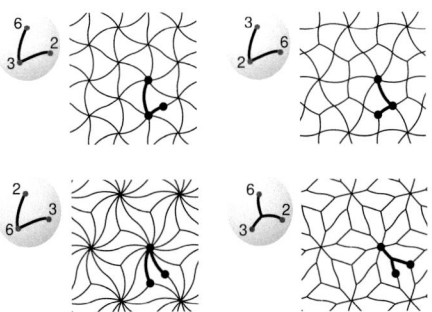

In fact, this must be so for every group with signature of the form **MNP**, with M, N and P distinct, since they will have topologically the same graphs; this holds for both spherical and planar symmetry types.

For signatures with (say) $M = N \neq P$, we have three possible labeled graphs, shown below for **332** and **442**. For those with $M = N = P$ we have just two, shown for **333**.

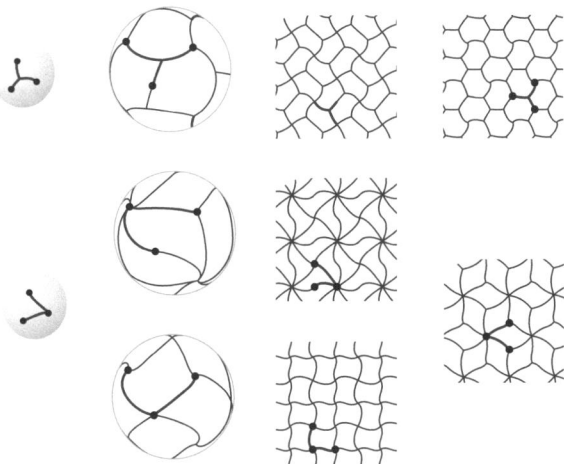

Earlier in this chapter, we saw that splitting open an envelope to lay flat produces a fundamental region for a pattern with signature **2222** — the envelope was the pattern's orbifold. We can work out that there are five different ways to draw a graph that splits open the orbifold, giving us the five Heesch types for this signature:

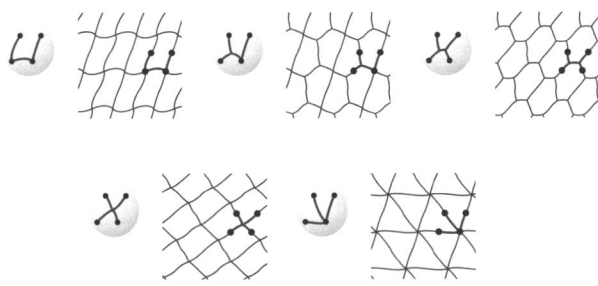

Symmetry types with with topologically equivalent orbifolds must have corresponding Heesch types since they will have topologically the same graphs. For example the orbifolds with signature **4∗2**, **3∗3**, **3∗2**, and **2∗N** are disks with a cone point in the interior and a kaleidoscopic point on the boundary. There are just two ways to draw a graph that splits such an orbifold into a fundamental region, and so there are two Heesch types for each of these signatures, as there are for any symmetry with a signature of the form **M**∗**N**.

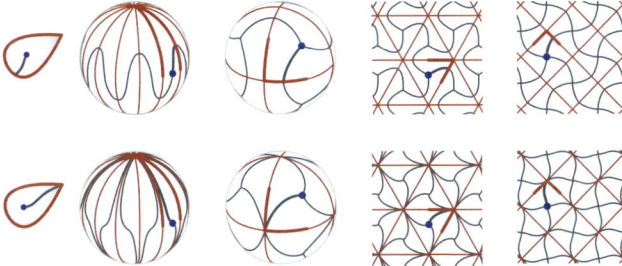

Because symmetries with signatures of the same typographical form have the same topology of orbifold, we may work out the Heesch types for all of them at once — whether in the Euclidean plane, on the sphere, or (Chapter 10) even in the hyperbolic plane — by enumerating graphs on this orbifold that split it into a disk and pass through all its cone points. You may wish to work out the remaining Heesch types for yourself before reading on!

Symmetry **22**∗ has four Heesch types, as would any symmetry of the form **NN**∗. (How many Heesch types would **MN**∗, $M \neq N$, have?)

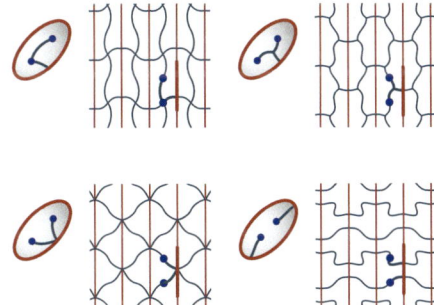

There is only one Heesch type with symmetry **N**∗, and there are two for **2**∗**22** (as there are for any of the form **M**∗**NN**).

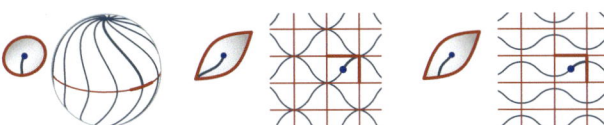

Orbifolds that are not simply connected are a little more challenging. Here are the Heesch types for **22**×:

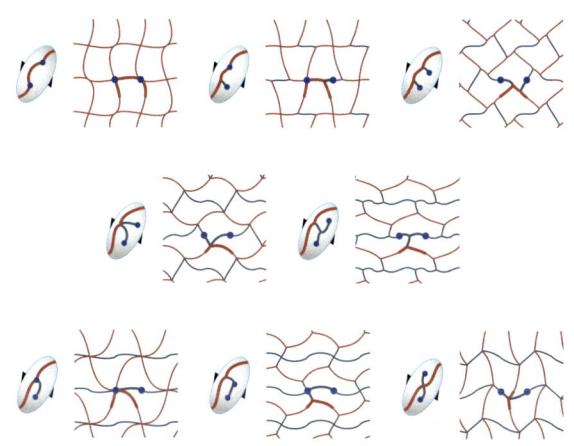

For ××:

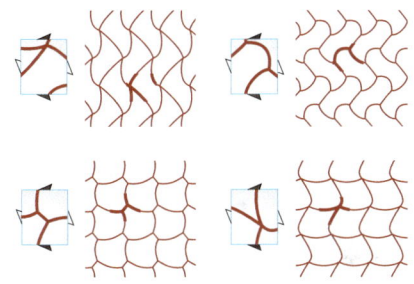

For ∗∗ and **N**×:

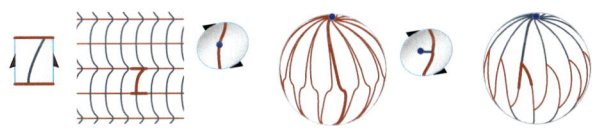

For ×∗:

Finally, there are two Heesch types for symmetry ○:

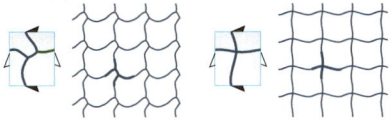

The Archimedean Polyhedra and Tilings

In Table 9.1 are the Archimedean tessellations of the sphere and Euclidean plane. These each have regular polygons for tiles and all vertices are of the same kind once we account for symmetry.

Consequently, on each tiling's orbifold, there is just one vertex (perhaps folded over or on a cone point). Each different kind of edge in a tiling corresponds to an edge on the orbifold (perhaps half an edge or a quarter edge, if it passes through a mirror or a gyration point).

In Chapter 19 of the full edition of *The Symmetries of Things* we show that each arrangement of edges determines the form of the orbifold and the typographical type of the symmetry. On the sides of these pages, we show the orbifolds for the tessellations beside them, with parameters a, b, etc. that shape the geometry of the pattern.

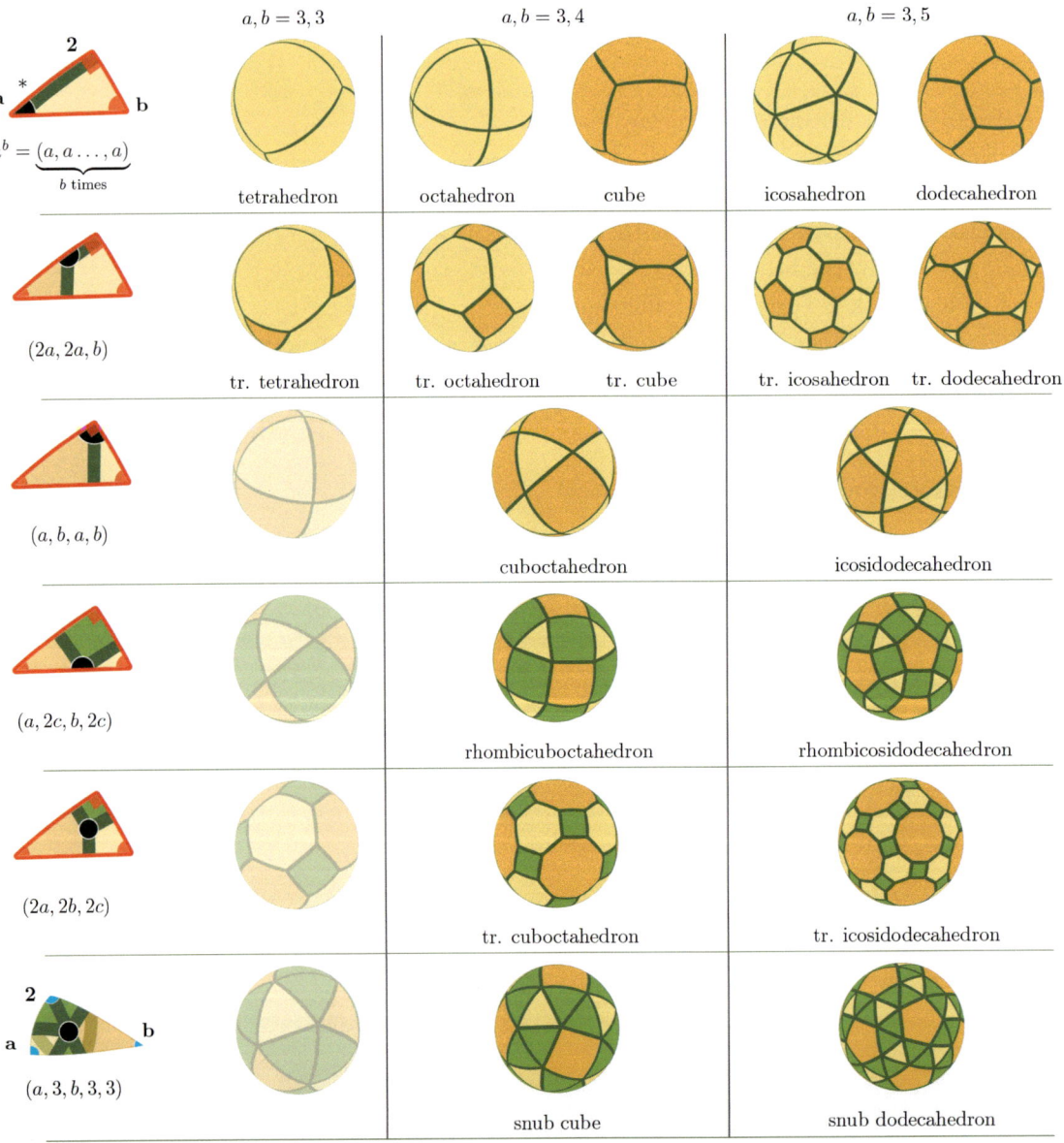

TABLE 9.1. *Regular and Archimedean polyhedra...*

For example in each of the patterns in the first row, each edge in the patterns has a 2-fold kaleidoscopic point at its center. On the corresponding orbifolds, there is a quarter edge. An orbifold for an Archimedean tiling with a quarter edge upon it can only be of the form *$*\mathbf{2ab}$. As we've determined which a and b yield spherical or plane repeating patterns, this enumerates the Archimedean tilings of this form.

Planar and spherical orbifolds have room only for a few edges at a vertex, but any arrangement is possible if we consider Archimedean tilings of the hyperbolic plane — a few examples appear on page 156.

In the table, we've abbreviated *truncated* as *tr* and "relative" Archimedean tilings are faded out. Such tilings have an additional symmetry and appear higher up in simpler form.

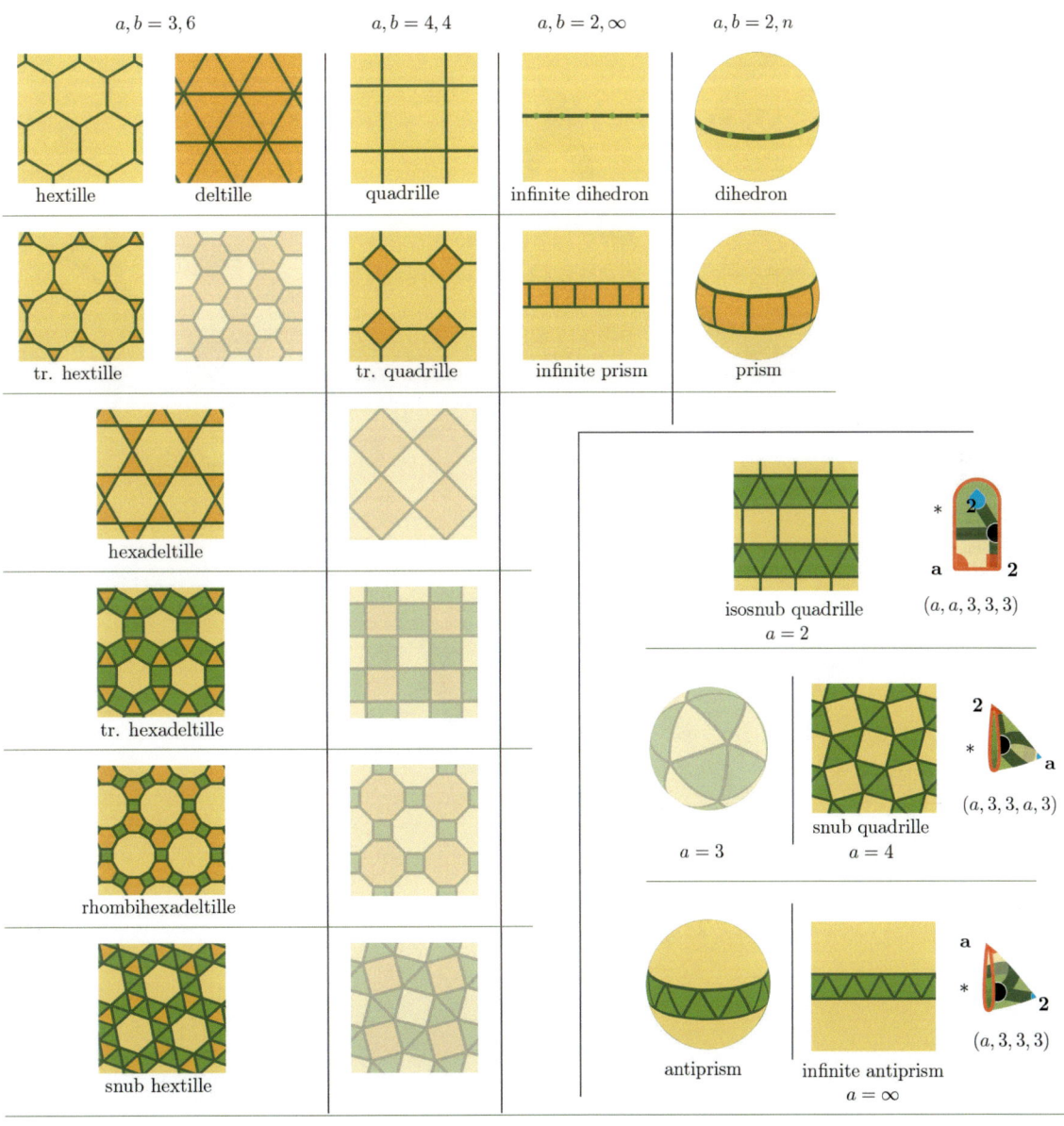

... and tilings, all in families according to the patterns of edges on their orbifolds.

Chapter 10

A Bigger Picture

In Chapter 3, we introduced the "cost" of each fundamental feature, and a magic theorem that classifies the Euclidean plane symmetry types — exactly those with signatures that cost exactly $^\$2$. In Chapter 4 we saw that signatures costing less than $^\$2$ perform the same service for the spherical groups.

What about signatures like ∗**732**, **23**×, or even **234**○×∗∗7∗**89**, that cost more than $^\$2$?

Remarkably, every one of these also describes a symmetry type, in the non-Euclidean geometry of the hyperbolic plane, discovered by Nikolai Lobachevski in 1829 and named by Arthur Cayley. William Thurston proved that in this geometry there is enough room to fit in as many features as we wish [17].

In fact he showed that *every* orbifold signature with the exceptions of those of the form **N**, **MN**, ∗**N**, or ∗**MN** with $M \neq N$, describes a symmetry type for repeating patterns — Euclidean if its cost is $^\$2$, spherical if less, hyperbolic if more. Moreover, these are exactly all of the two-dimensional symmetry types.

In the full edition of *The Symmetries of Things* we give a more complete account, but you won't need to know any hyperbolic geometry to read this book because we can say it all in pictures. You can make your own patterns with tools like the *Kaleidesign* software that we use to create these illustrations, and you can even experience life in non-Euclidean worlds by playing video games like the ones we showcase on pages 160 and 161!

(Opposite page and above) In the hyperbolic plane, all of these daffodil stems are the same size and shape. In this pattern, look for the 5-fold, 4-fold, and 3-fold gyration points, and verify that there one of each kind — the pattern has symmetry type **543**. In order to draw it flat in the pages of this book we had to distort its geometry, applying the Poincaré disk projection, opposite, and the Klein projection, above.

We have already encountered some non-Euclidean geometry — that of the sphere. In the Euclidean plane, the Parallel Principle says there is just one line parallel to a given one through a given point, but on the sphere every "straight line" (i.e great circle) meets every other, and there are no parallel lines anywhere at all. The sphere is more constricted than is the Euclidean plane.

In hyperbolic geometry, the opposite applies, and there are infinitely many lines through a point that don't meet a given one. This means that hyperbolic space is vastly more expansive than the Euclidean plane.

The fact that the hyperbolic plane is not Euclidean does not stop us from drawing pictures of it, just as nothing prevents us from drawing distorted pictures of a sphere — we are used to images like the one at right, of a spherical pattern with symmetry type **532**.

The pattern below is just the same, only projected from the sphere in a different manner. Its type is also **532** — even when they are distorted these patterns remain spherical, and their orbifolds are equivalent.

*A spherical pattern of type **532**.*

*The same spherical pattern of type **532**.*

We have no trouble viewing a Euclidean pattern on a piece of fabric or wrapped around the side of a vase, and we can distort a picture of a planar pattern without changing its type. In the illustration at right, if we imagine all of the flashlights are the same shape and size, we find 6-fold, 3-fold, and 2-fold gyration points, and only one of each kind. The symmetry type is **632**, and this pattern is actually Euclidean!

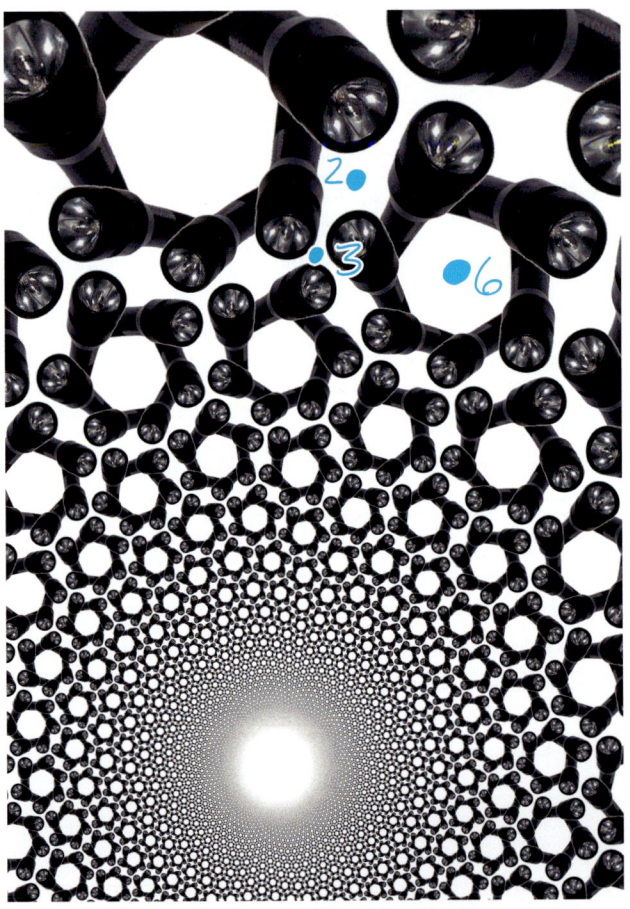

*A Euclidean pattern of type **632**.*

There are countless ways to render the hyperbolic plane on a flat page. The two most common 'projections' of the hyperbolic plane take it to a disk in the Euclidean one. One of them, found by Beltrami and named for Klein [15], shown at top left below, takes hyperbolic straight lines to (segments of) Euclidean ones. This is really the most natural projection, because it is in fact the way the hyperbolic plane would appear if you viewed it from a point in hyperbolic 3-space, as in the first-person video games shown on page 161.

However, the other common method, usually credited to Poincaré though earlier known to Beltrami [15], top right below, is more widely known to both mathematicians and artists. It is the projection used in Maurits C. Escher's famous "Circle Limit" engravings and is the one we use for most of the illustrations in this chapter. It preserves angles but takes hyperbolic straight lines to arcs of circles perpendicular to the boundary. The reason this less-natural projection is so often used is that it shows more of the plane. It is equivalent by inversion to the upper half-plane model often used by mathematicians.

But there are infinitely many possibilities — on the bottom row, we show this pattern with signature **433** in two more projections that also preserve angles.

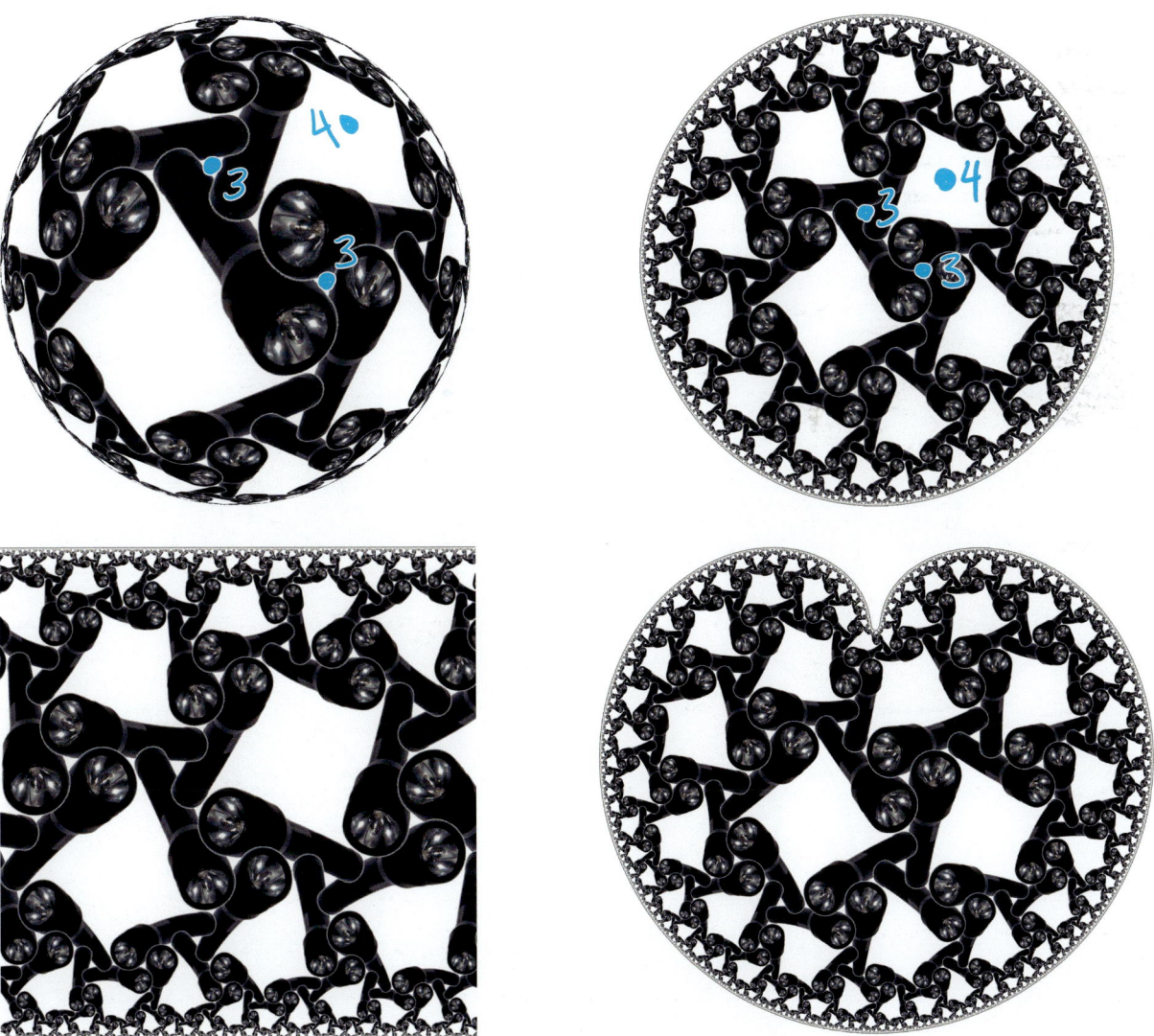

Different projections of the same pattern, with signature **433**.

On the previous pages, we saw several patterns that looked somewhat the same although they had very different geometries. They each had signatures of the form **PQR**, and orbifolds of the same topological type, a sphere with three cone points. In the same way, with the exceptions of **NN** and ∗**NN**, any repeating symmetry type with a digit in its signature generalizes to infinitely many types in the hyperbolic plane.

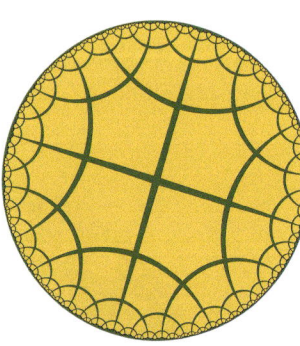

Although there are only five regular polyhedra and three regular planar tilings (listed on the first row of Table 9.1), if we include regular tilings of the hyperbolic plane, then M-gons may meet N-to-a-vertex (denoted $\{M, N\}$) for any $N, M \geq 3$, with signature ∗**2NM**. Above is a drawing of a $\{5, 4\}$ tiling by right-angled pentagons — in the hyperbolic plane these are all the same shape and size.

Each row of Table 9.1, showing a type of Archimedean tiling defined by a topological arrangement of edges on an orbifold, is only the beginning of an infinite list of tilings differing only in the orders of their gyration and kaleidoscopic points. For example, below we see the rhombi-$\{5, 4\}$ and the snub $\{5, 4\}$, which properly belong with the tilings on the last two rows of the table with $a = 5$ and $b = 4$.

The bottom two tilings are more novel. The tiling at left has symmetry type **3322** if we pay attention to the colors. If we ignore the colors, it has additional symmetries and its type is **32×** — a fundamental domain for that type is outlined. At right, the colors do not affect the symmetry and either way the symmetry type is **2∗43**. (The corresponding Euclidean tilings in **22×** and **2∗22** are by squares, meeting four-to-a-vertex. Like the shaded out entries in the table, these are only "relative" — they have further symmetry and a simpler orbifold.)

We give a method for enumerating all Archimedean tilings Chapter 19 of *The Symmetries of Things*, and still more exotic tilings appear at the end of this chapter.

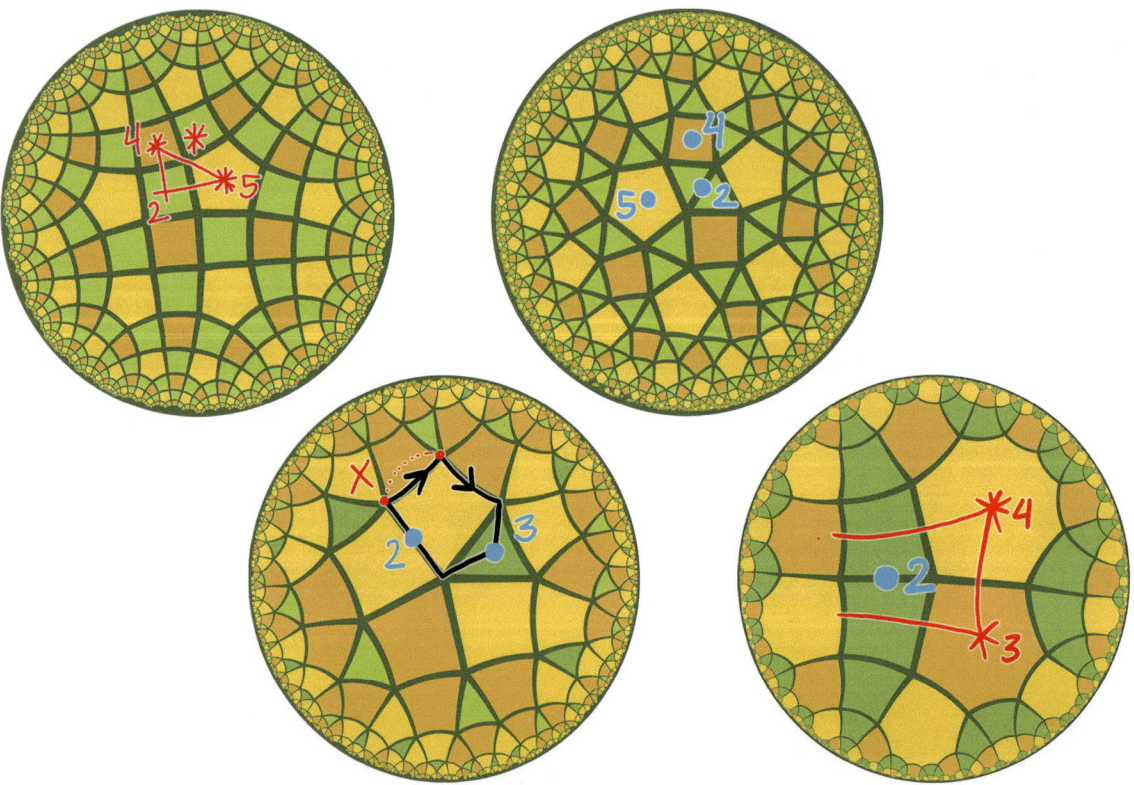

Archimedean tilings in the hyperbolic plane.

A physical model of the hyperbolic plane.

Hyperbolic geometry can fit into the real world on surfaces that are saddle-shaped everywhere, like this sculpture is.

The blue strips form regular polygons, with equal edge lengths and angles, if we measure them within the surface defined by the sculpture. Three squares and a regular pentagon fit together at each vertex of this pattern. This is a piece of an undistorted rhombi-{5,4} tiling, placed into our space!

The yellow paths are straight in this surface, as they must be because steel strips cannot easily be bent to veer sideways. They lie along most of the mirror lines of the pattern — can you spot the others? — and the pattern has signature ∗**542**.

In fact, everything that we have done in this book for spherical, Euclidean, and frieze types extends easily to the hyperbolic case. This is because we have been working with the orbifolds that underly these patterns, and these orbifolds are fundamentally topological in nature. They are compact connected surfaces with specially marked points — cone points in their interiors and kaleidoscopic corner points on their boundaries.

The geometry of an orbifold and the patterns that it produces are determined by its topological type and the orders of these special points. Therefore, anything that is unchanged when we relabel the orders of our special cone and corner points will remain the same across a range of patterns, those with the same typographical type, such as **M∗N** or **PQR**.

In the full edition of *The Symmetries of Things*, we give other applications of this principle. For example, to produce the table of Archimedean tilings and polyhedra at the end of Chapter 9, we enumerated the ways that edges of the tiling may be placed on a spherical or planar orbifold. Any symmetry with the same topological type of orbifold will have corresponding tilings, like the ones on the previous page. Moreover, the same general method can be continued indefinitely, listing out new kinds of Archimedean tilings in the hyperbolic plane with larger and larger orbifolds.

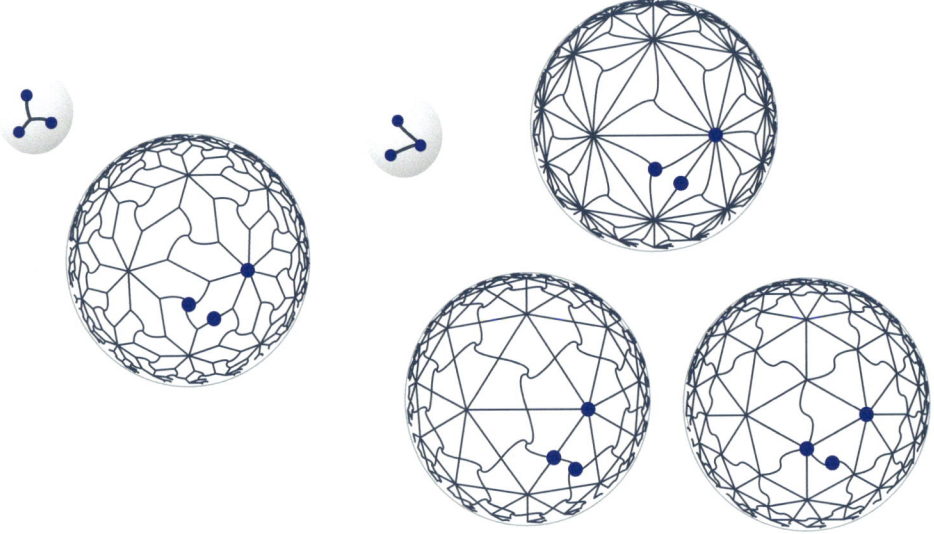

(Above) Heesch types with symmetry **732**.

On pages 148 and 149, we enumerated the Heesch types for planar and spherical symmetries by looking at graphs on their underlying orbifolds. Just as there are four Heesch types of symmetry **632**, there are four corresponding types for **732**, shown above, as there are for any **PQR** with distinct P, Q, and R.

Similarly, at right are several patterns of signature **M∗N**. Each has a disk as its orbifold, with a path connecting a cone point of order M in its interior to a kaleidoscopic point of order N on its boundary ∗. These points may be of infinite order, as with the frieze pattern on the far right on the top row.

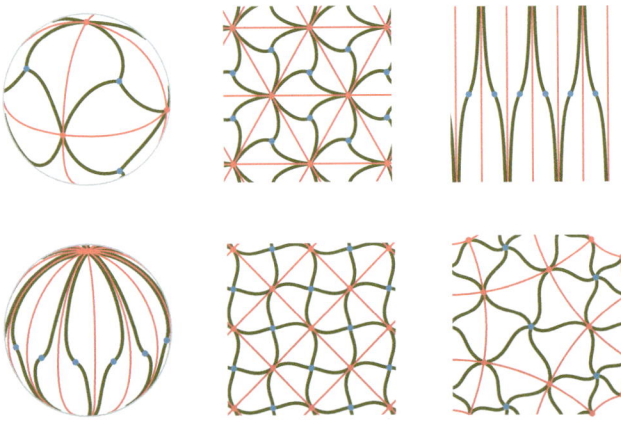

Heesch types with topologically equivalent orbifolds.

Where Are We?

In this chapter we took a wider view: If we include the patterns on the hyperbolic plane, every orbifold signature (with a few simple exceptions) describes a symmetry type in *some* geometry.

In Chapters 1 through 8 we have proven this for signatures costing $2 or less, but we only hinted what this means for those costing more than $2.

In this book, we haven't explained how the typical hyperbolic planar symmetry type may be isotoped and changed continuously, like the duck designs on this page have been,

or indeed that geometric symmetry groups with the same signature must always be isotopic. In a complete accounting, we would explain that the area of a fundamental region is related to its cost, and we would have a lot to add about hyperbolic geometry itself.

This awaits a future richly illustrated work. In the meantime, you can turn to Chapters 17 and 18 of the full edition of *The Symmetries of Things* to learn more, or experiment with our new *Kaleidesign* software and make beautiful illustrations of your own.

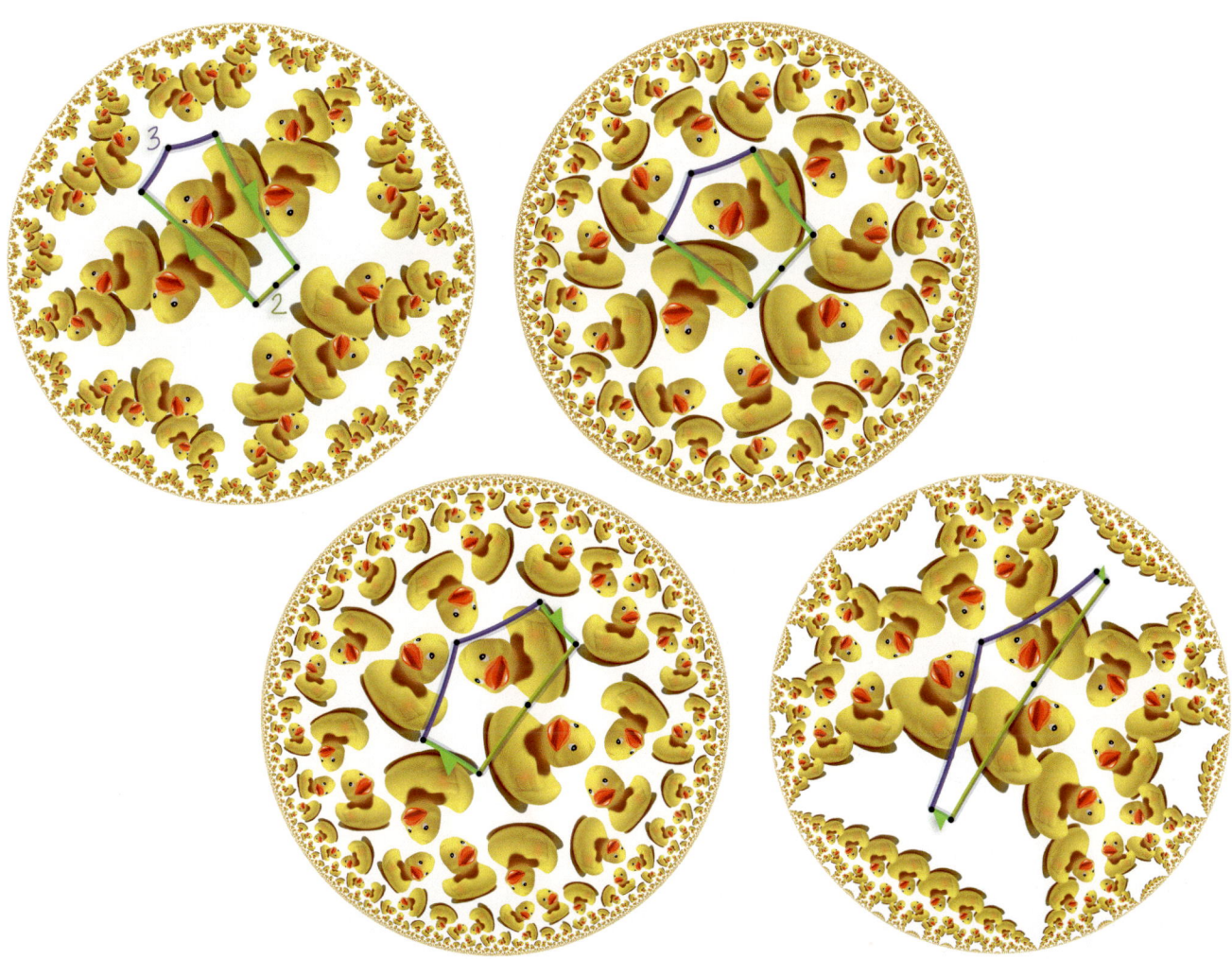

Isotoping patterns of symmetry type **23**×.

Non-Euclidean Video Games

The best way to gain a feel for hyperbolic geometry is to work and play there. Now we can! Here are some of our favorite games set within non-Euclidean worlds. In *Hyperrogue*, by Eryk Kopczyński and Dorota Celińska, you can adventure, slay monsters and explore strange lands [7]. The game is set is set on a "hyperbolic soccer-ball", a.k.a. the truncated {7,3}, an Archimedean tiling with heptagons surrounded by hexagons. There is much more room around you than in the Euclidean plane — you soon realize that the hyperbolic plane is truly capacious.

Each of this game's puzzles reveals another feature of hyperbolic geometry, and within its settings is an encyclopedia of projections, other geometries, and tilings.

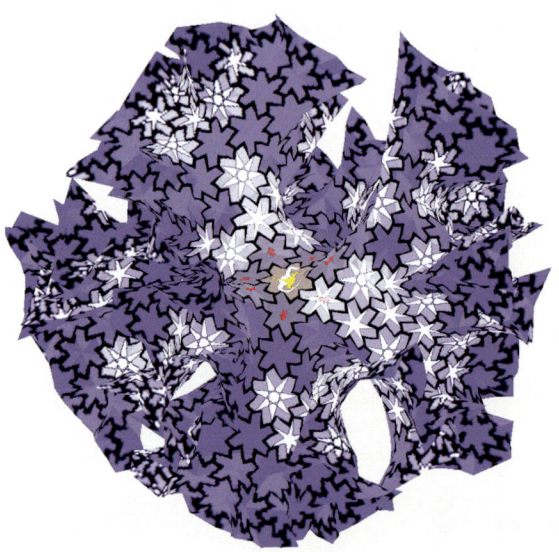 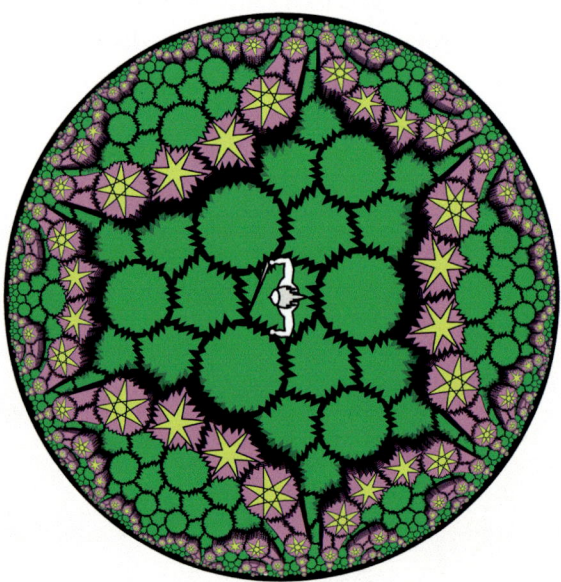

Hyperrogue *in different models of the hyperbolic plane.*

On our phones we play *Non-Euclidean Minesweeper* on a regular tiling of our choice (left) [12]. Jeff Weeks' *Hyperbolic Games* (middle and right) [19] are played on an underlying grid with ∗**732** symmetry, but the boards themselves are fundamental regions for a pattern of type ○○○! (This pair of symmetries shows the famous "Klein Quartic.")

Non-Euclidean Minesweeper.

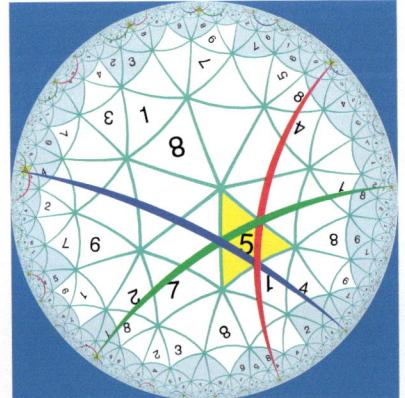

Sudoku.

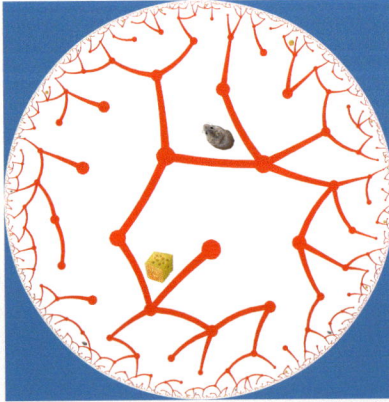

A Maze.

Immersing ourselves in Jeff Weeks' *Curved-Space Pool Hall* we strike balls on a non-Euclidean table, with five straight sides and five right-angled corners [20]. In any geometry we inhabit, angles will be distorted unless we look directly down upon them, but as you can see, straight paths will appear straight to our eyes. This is even so in the curved geometry of the hypersphere, one of the many geometries to visit in CodeParade's *Hyperbolica*, shown below [2].

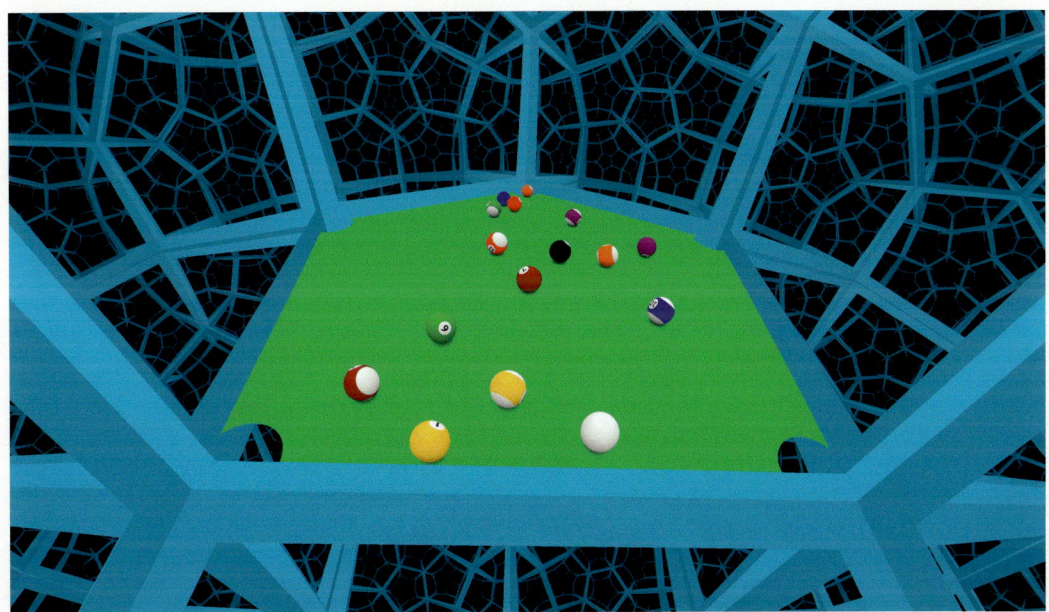

Hyperbolic billiards on a right-angled pentagon.

Life in spherical geometry.

Examples and Exercises

These patterns have only gyration points.
What are their signatures? Answers are on page 165.

What are the symmetry types of these patterns? (The patterns are
each distorted on this page, but they aren't all necessarily non-
Euclidean...)

The pattern at top left has a signature of the form **M∗N**. Can you identify M and N? The orbifolds of the other patterns on this page have topological features. You can work out each signature by finding a fundamental region, and then examining how it attaches to copies of itself — by attaching the fundamental region to itself in the same way we form the pattern's orbifold. We can use the tools we learned about in Chapter 8 to find the orbifold's topology, and hence the symmetry type.

For example, we've outlined a fundamental region in the upper right image. After identifying like edges, we find the orbifold has two cone points, of orders 2 and 3, and the topology of a crosscap. The symmetry type is **23×**.

The figure below left has two kinds of mirror lines, that don't intersect, and one kind of gyration point. What is its signature? You can verify that it is **3∗∗** by checking its orbifold is an annulus with a single cone point.

We've marked a fundamental region for the pattern below right. On the orbifold the markings form a graph with one face, three vertices, and five edges. The pattern has orientation-reversing symmetries. The orbifold is therefore non-orientable, with Euler characteristic -1. The symmetry type of the pattern and the topological type of its orbifold are thus both ○× = ×××.

We conclude for now with a few of the many thousands of tilings by regular polygons that Marek Čtrnáct has enumerated in the hyperbolic plane [4]. The top two are not uniform, in the sense that they have more than one kind of vertex arrangement within them. The bottom two have identical polygons, arranged differently. Can you identify the symmetry types of these patterns?

Answers

On page 162 we have at top **433** and **543**, **732** in the center, and **(20)32** and ∞**32** at bottom. On page 163 appear **732** (in the Klein model of the hyperbolic plane), **532** (in a picture of a sphere), and **632** (in an unusual projection of the Euclidean plane). The pattern at upper left on this page has symmetry type **32**∗. At upper right, each polygon has a gyration point at its center, and the type is **3332**. The pattern at lower right has signature **4**∗**3**. The pattern at lower left is trickier — the blue and the green squares cannot be interchanged, but polygons of the same color can be. With mirrors along the rows of green squares, the type is **3222**∗.

Appendix A

Other Notations for the Plane and Spherical Groups

The columns of Table A.1 correspond to different notation systems, subject to the remarks noted below. The column titles are abbreviations:

OS	our orbifold signature
I	International notation
C&M	Coxeter and Moser
S	Speiser
N	Niggli
P	Pólya
G	Guggenheim
F	Fejes Tóth
C	Cadwell

Our orbifold signature is the one presented in this book [1]. The International notation is the most used of the older notations. The C&M notation is the notation used in Coxeter and Moser's *Generators and Relations for Discrete Groups* [3], which should be consulted for the individual references.

The notations p3m1 and p1m3 were inadvertently interchanged by Niggli, whose notation is otherwise taken from Spieser with the addition of the Roman numerals in parentheses. This error is repeated in editions of Coxeter and Moser before 1980, by which time Doris Schattschneider [10] and H. Martyn Cundy [5] had independently discovered the interchange, and in many other places. We thank Schattschneider for this information.

Table A.2 compares our signature with older notations for the spherical groups; it is adapted from Coxeter and Moser's *Generators and Relations for Discrete Groups*, which should again be consulted for the references. The reader should be warned that the fonts have been uniformized for simplicity and that for N = 1 or 2 there are various special notations and equivalences that we have ignored, since they become obvious from the signature when digits 1 are omitted. Some pairs of lines contain notations in braces, which as they stand are for even values of N, but should be interchanged when N is odd. The abbreviations for Table A.2 are as follows:

OS	our orbifold signature
C	Coxeter
S	Schoenflies
W	Weyl
P&M	Pólya and Meyer
I	International notation

OS	I (C&M)	S (N)	P G	F C
*632	p6m	$C_{6v}^{(I)}$	D_6	W_6^1
632	p6	$C_6^{(I)}$	C_6	W_6
*442	p4m	$C_4^{(I)}$	D_4^*	W_4^1
4*2	p4g	C_{4v}^{II}	D_4°	W_4^2
442	p4	$C_4^{(I)}$	C_4	W_6
*333	$\left(\begin{matrix}\text{p3m1}\\\text{p31m}\end{matrix}\right)$	$\left(\begin{matrix}C_{3v}^{II}\\C_{3v}^{I}\end{matrix}\right)$	D_3^*	W_3^1
3*3			D_3°	W_3^2
333	p3	$C_3^{(I)}$	C_3	W_3
*2222	pmm	C_{2v}^{I}	$D_2 kkkk$	W_2^2
2*22	cmm	C_{2v}^{IV}	$D_2 kgkg$	W_2^1
22*	pmg	C_{2v}^{III}	$D_2 kkgg$	W_2^3
22×	pgg	C_{2v}^{II}	$D_2 gggg$	W_2^4
2222	p2	$C_2^{(I)}$	C_2	W_2
**	pm	C_s^{I}	$D_1 kk$	W_1^2
*×	cm	C_s^{III}	$D_1 kg$	W_1^1
××	pg	C_2^{II}	$D_1 gg$	W_1^3
○	p1	$C_1^{(I)}$	C_1	W_1

TABLE A.1. *The Euclidean plane groups.*

OS	C	S	W	P&M	I
*532	[3,5]	I_h	$\bar{P}$	I	53m
532	$[3,5]^+$	I	P	I	532
*432	[3,4]	O_h	$\bar{W}$	O_i	m3m
432	$[3,4]^+$	O	W	O	432
*332	[3,3]	T_d	WT	T_O	$\bar{4}3m$
3*2	$[3^+,4]$	T_h	$\bar{T}$	T_i	m3
332	$[3,3]^+$	T	T	T	23
*22N	[2,N]	D_{Nh}	$\left(\begin{matrix}\bar{D}_N\\D_{2N}D_N\end{matrix}\right)$	$\left(\begin{matrix}D_N i\\D_N\,D_{2N}\end{matrix}\right)$	N/mmm or $2\bar{N}$ m2
2*N	$[2^+,2N]$	D_{Nd}			$2\bar{N}$ 2m or $\bar{N}$ m
22N	$[2,N]^+$	D_N	D_N	D_N	N2
*NN	[N]	C_{Nv}	$D_N\,C_N$	$C_N\,D_N$	Nm
N*	$[2,N^+]$	C_{Nh}	$\left(\begin{matrix}\bar{C}_N\\C_{2N}C_N\end{matrix}\right)$	$\left(\begin{matrix}C_N i\\C_N\,C_{2N}\end{matrix}\right)$	N/m or $2\bar{N}$
N×	$[2^+,2N^+]$	S_{2N}			$2\bar{N}$ or $\bar{N}$
NN	$[N]^+$	C_N	C_N	C_N	N

TABLE A.2. *The spherical groups.*

Additional Teaching Materials

These materials and more appear at
themagictheorem.com

Please download them and enjoy! For further intuition, we highly recommend playing with a symmetrical drawing program, such as Jeff Weeks' *KaleidoPaint*, Jürgen Richter-Gerharts' *iOrnament*, or our own *Kaleidesign*.

Illustration Credits

Throughout this book, the symmetrization of photographic imagery is inspired by Chris Whatley's *Tess*, written for the NeXT computer, ca. 1992. We used software written with the assistance and collaboration of many people, especially Troy Gilbert and Vladimir Bulatov.

Page 2: Gryphons, Chicago, photograph by Susan McBurney. Tracery drawing from *Maßwerk*, Günther Binding (Wissenschaftliche Buchgesellschaft, Darmstadt, 1988), Kościół Mariacki, Toruń (Thorn). Tracery photographs: Central Park, New York.

Page 4: Snakes: *1,100 Designs and Motifs from Historical Sources*, John Leighton (Dover Press, New York, 1995). Tracery: from *Maßwerk*, Bazylika Katedralna, Pelplinie (Pelplin); Frauenkirche, Esslingen.

Page 5: From *Maßwerk*: (top) St. Peter und Paul, Leignitz; Bazylika Katedralna, Pelplinie (Pelplin); Peterskirche, Görlitz; Kathedrale Notre-Dame, Riems; Jakobikirche, Neiße. (bottom) Pelplinie (Pelplin); and Kościół Mariacki, Toruń (Thorn).

Page 6: Rosettes: seen in New York City.

Page 7: Bow Bridge, New York; Medinah Temple, Chicago (photo: Susan McBurney); New York; CGS; New York; cylinder seal: Ashmolean Museum, Oxford AN1949.900. (photo: Zunkir, *Wikimedia Commons*).

Page 8: Shirt: designed by Jhane Barnes, modeled by Stan Isaacs; Bethesda Terrace; midtown; Bloomingdales; Bethesda Terrace; MoMath after N. Myrhvold, all in New York City.

Page 9: Dice Labs; Eigil Nielsen (photo: unknown); Carolyn Yackel (photo: C.Y.); CGS, found using Robert Webb's *Great Stella* software; Dick Esterle (photo: D.E.); Bathsheba Grossman (photo:

B.G.); Jack Puzzle by Craighill (photo C.H.); John Kostick; Shiying Dong (photo S.D.).

Page 30: *La Mano* lotería card, unknown.

Page 45: Top row: New York City; R. Guastavino, Queensboro Bridge, NYC; Holiday Inn Express, Abilene. Second row: NYC; NYC DoT; Abilene. Third row: Bristol; Abilene; USA.

Page 53: Top row: Marine Air Terminal, LGA, New York. Imperial College, London; Brooklyn. Second row: Terminal 2, LGA; NYC; NYC. Third Row: photographed in Olympia, widely seen; NYC; NYC, widely known.

Page 55: Top: London; Cassis; London. Middle: Ljubljana business district; Cassis; unknown. Bottom: Roosevelt Island, NYC; UN, NYC; Aberyswyth.

Page 69: Eyeglass idea: unknown. Soccerball Pov-ray file, Remco de Korte.

Page 70: (center top) Shiying Dong; (lower right) Judy Peng.

Page 71: Stellations found with *Great Stella* software, by Robert Webb. Painted compound of three cubes: Zoe Curlee-Strauss.

Page 72: Bathsheba Grossman (photos: B.G.).

Page 73: Bathsheba Grossman (photos: B.G.).

Page 74: Lower right, unknown; others, Ginny Thompson.

Page 75: Carolyn Yackel (photos: C.Y.).

Page 76: Jon-Paul Wheatley (photos: J.-P.W.).

Page 77: Collected by David Swart; see [16].

Page 78: (Lower left) Elsa Pandozi (photo: E.P.).

Page 79: (Lower left and upper right) Elsa Pandozi (photo: E.P.).

Page 80: *Kaleidotile*, Jeff Weeks.

Page 80: Five-, seven-, nine-, eleven-, fourteen-, eighteen- and twenty-two-sided dice, Impact Miniatures; the red fourteen-sided d7, and black tetrakis hexahedron, Koplow; others, Dice Lab.

Page 91: Friezes: first four, Chicago (photo: Susan McBurney); latter three, New York.

Page 103: After Robert Dixon, *Mathographics* (Dover Press, New York, 1991).

Page 106: Image and software, Ken Stephenson; combinatorics: Jim Cannon, Bill Floyd, and Walter Perry.

Page 114: Shiying Dong (photo: S.D.).

Page 115: Shiying Dong (photo: S.D.).

Page 126 and thoughout, after George Francis.

Page 137: *Tessellation Station*, National Museum of Mathematics. Tiles: Makoto Nakamura. Orbifold idea: Jeffrey Wack.

Page 159: Duck: Tub Time L'il Duck.

Page 160: *Hyperogue*: Eryk Kopczyński, Dorota Celińska and Marek Čtrnáct [7]. *Non-Euclidean Minesweeper*: Sci-Tech Binary, Ltd. Co. [12]. *Hyperbolic Sudoku*: Jeff Weeks [19].

Page 161: *Hyperbolica*: CodeParade [2]. *Curved Space Pool Hall*: Jeff Weeks [20].

Page 163: Boat: unknown.

Page 164: *La Mano* lotería card, unknown.

Page 165: Marek Čtrnáct.

All other illustrations, models, and photographs are by Chaim Goodman-Strauss.

Bibliography

[1] John H. Conway. "The Orbifold Notation for Surface Groups." In M. W. Liebeck and J. Saxl, editors, *Groups, Combinatorics and Geometry: Proceedings of the L.M.S. Durham Symposium, July 5, Durham, U.K., 1990*, London Mathematical Society Lecture Note Series 165, pages 438–447, Cambridge, 1992. Cambridge University Press.

[2] CodeParade *Hyperbolica*. [Microsoft Windows, Linux, macOS app] Steam, 2022.

[3] H.S.M. Coxeter and W. Moser. *Generators and Relations for Discrete Groups*. Ergebnisse der Matematik und Ihrer Grenzgebrete 14. Springer, Berlin, 1957. (Later editions published in 1965, 1972, and 1980.)

[4] Marek Čtrnáct. *https://zenorogue.github.io/tes-catalog*

[5] H. Martyn Cundy. "p3m1 or p31m?" *Mathematical Gazette*, 63:192, 1979.

[6] P. Diaconis and J.B. Keller. "Fair Dice." *American Mathematical Monthly*, 96:337-339 1989.

[7] Eryk Kopczyński, Dorota Celińska, and Marek Čtrnáct. "HyperRogue: Playing with Hyperbolic Geometry." *Bridges Conference Proceedings*, 2017.

[8] Branko Grünbaum and Geoffrey Shephard. *Tilings and Patterns*. W.H. Freeman & Company, New York, 1986.

[9] Ed Pegg. "Fair Dice." *www.mathpuzzle.com/Fairdice.htm*

[10] Doris Schattschneider. "The Plane Symmetry Groups, Their Recognition, and Notation." *American Mathematical Monthly*, 85:439–450, 1978.

[11] Doris Schattschneider. *M.C. Escher: Visions of Symmetry*. Harry N. Abrams, New York, 2004.

[12] Sci-Tech Binary *Non-Euclidean Minesweeper*. [iOs, Windows, and Linux app] Steam, 2024.

[13] Brigitte Servatius. "The Geometry of Folding Paper Dolls." *The Mathematical Gazette*, 81:29-36, 1997.

[14] Bobby Stecher and Carolyn Yackel. " Dyeing to Make an Orbifold." *Bridges Conference Proceedings*, 2024.

[15] John Stillwell. *Sources of Hyperbolic Geometry*. American Mathematical Society, Providence, 1999.

[16] David Swart. "Soccer Ball Symmetry." *Proceedings of Bridges*, pages 151–158, 2015. Tessellations Publishing, Phoenix.

[17] W. Thurston. Chapter 13 of *The Geometry and Topology of Three Manifolds. Electronic version 1.1 - March 2002*, Simons Laufer Mathematical Sciences Institute, *https://library.slmath.org/books/gt3m/*

[18] J. R. Weeks and G. K. Francis. "Conway's Zip Proof." *American Mathematical Monthly*, 106:393–399, 1999.

[19] Jeff Weeks. *Hyperbolic Games*. [iOs app] App Store, 2020.

[20] Jeff Weeks. *Curved-Space Pool Hall*. [visionOs app] App Store, 2025.

Index